35·00

The JCT Minor Works Form of Contract

Second E

David
and
Vincen

b

Blackwell Science

Blackwell Science Ltd
Editorial Offices:
Osney Mead, Oxford OX2 0EL
25 John Street, London WC1N 2BL
23 Ainslie Place, Edinburgh EH3 6AJ
350 Main Street, Malden
    MA 02148 5018, USA
54 University Street, Carlton
    Victoria 3053, Australia
10, rue Casimir Delavigne
    75006 Paris, France

Other Editorial Offices:

Blackwell Wissenschafts-Verlag GmbH
Kurfürstendamm 57
10707 Berlin
Germany

Blackwell Science KK
MG Kodenmacho Building
7–10 Kodenmacho Nihombashi
Chuo-ku, Tokyo 104, Japan

First edition published 1990 by Legal Studies and
Services (Publishing) Ltd
Second edition published 1999 by Blackwell
Science Ltd
Reprinted 1999

Set in 10.5/12.5 pt Palatino
by DP Photosetting, Aylesbury, Bucks
Printed and bound in Great Britain by
MPG Books Ltd, Bodmin, Cornwall

DISTRIBUTORS

Marston Book Services Ltd
PO Box 269
Abingdon
Oxon OX14 4YN
(Orders: Tel: 01235 465500
          Fax: 01235 465555)

USA
Blackwell Science, Inc.
Commerce Place
350 Main Street
Malden, MA 02148 5018
(Orders: Tel: 800 759 6102
                781 388 8250
          Fax: 781 388 8255)

Canada
Login Brothers Book Company
324 Saulteaux Crescent
Winnipeg, Manitoba R3J 3T2
(Orders: Tel: 204 837-2987
          Fax: 204 837-3116)

Australia
Blackwell Science Pty Ltd
54 University Street
Carlton, Victoria 3053
(Orders: Tel: 03 9347 0300
          Fax: 03 9347 5001)

A catalogue record for this title
is available from the British Library

ISBN 0-632-03967-1

Library of Congress
Cataloging-in-Publication Data
is available

For further information on Blackwell Science, visit
our website: www.blackwell-science.com

Contents

Contents

Contents

Preface to the Second Edition

The Agreement for Minor Building Works (which is still referred to for convenience as MW 80) is the most widely used of the JCT forms of contract. Since 1968, when it was first issued, it has been used consistently not only for simple short contracts of moderate price for which it is intended, but also for much larger projects for which it is often not suited at all. It is a very short form and no doubt its brevity accounts for much of its popularity.

Its simplicity, however, is deceptive. That which is omitted may be as significant as the express terms. In the preface to the first edition, we remarked that, like all forms of contract, it must be read against the considerable and growing body of case law. Unfortunately, the case law continues to grow with no equivalent dropping off of cases which are no longer relevant.

This book explains the practical applications of the form from the points of view of the employer, the architect and the contractor. It is not a clause-by-clause analysis. The form has been considered under topic headings where the effect of various clauses can be gathered together. Legal language has been eschewed in favour of simple explanations of legal concepts supported by flowcharts, tables and sample letters. Where possible, we have attempted to provide a clear statement of the legal position in a variety of common circumstances together with references to decided cases so that those interested can read further. Occasionally, where the precise legal position is not clear, a view has been offered. This book is intended to be a practical working tool for all those using MW 80.

Almost a decade has passed since the first edition was drafted. During that time, much has changed. This edition has been heavily revised to take account of a substantial number of relevant new cases and the textual reference to cases has been almost doubled to respond to the clear demand for this kind of information. All the JCT amendments have been taken into account including Amendment No. MW11 of April 1998, relevant changes in legislation including principally the Housing Grants, Construction and Regeneration Act 1996 and the associated Scheme for Construction Contracts (England and Wales) Regulations 1998, the Arbitration

Act 1996, the Construction (Design and Management) Regulations 1994, the Companies Act 1989, the Law of Property (Miscellaneous Provisions) Act 1989 and the RIBA terms of engagement commonly known as SFA/92, CE/95 and SW/96.

The untimely death of Professor Vincent Powell-Smith has resulted in the burden of revision falling entirely on the remaining author. It is a tribute to Vincent's formidable expertise, for which I am profoundly grateful, that only one revision has been generated solely by a mature consideration of the original facts.

Fortunately, I have been able to persuade my colleague Derek Marshall LLB (Hons), MASI, ACIArb, to completely revise the chapter on arbitration to deal with the new Act and to cover the basics of adjudication, drawing on his considerable experience in this field.

David Chappell
Chappell-Marshall Limited
27 Westgate
Tadcaster
North Yorkshire
LS24 9JB

CHAPTER ONE
THE PURPOSE AND USE OF MW 80

1.1 The background

The JCT Agreement for Minor Works was first published in 1968; it was revised in 1977. The headnote explained that it was intended for minor building works or maintenance work, based on a specification or a specification and drawings, to be carried out for a lump sum. It was inappropriate for use with bills of quantities or a schedule of rates.

Despite its shortcomings, it was widely used for small projects and even for larger ones, its main attraction being its brevity and apparent simplicity. The evidence suggests that architects were becoming increasingly dissatisfied with the length and complexity of the then current main JCT standard form contract (JCT 63) and then – as now – wished to use simple contract conditions wherever possible.

The Minor Works form was extensively revised by a JCT working party in 1979 and a new edition was published in January 1980 and reprinted with corrections in October 1981. It was revised again in April 1985, January 1987, March and October 1988, September 1989, January 1992 and March 1994. Amendment MW9:1995 was issued to take account of the Construction (Design and Management) Regulations 1994 and Amendment MW10:1996 dealt with more insurance changes, Amendment MW11:1998 makes significant changes arising out of the Latham Report and the Housing Grants, Construction and Regeneration Act 1996. In effect, MW 80 in its revised form is a completely new set of contract conditions. The headnote to the form as issued in 1980 set out its purpose:

'The Form of Agreement and Conditions is designed for use where minor building works are to be carried out for an agreed lump sum and where an Architect or Contract Administrator has been appointed on behalf of the Employer. The Form is not for use for works for which bills of quantities have been prepared, or where the Employer wishes to nominate sub-contractors or

1

suppliers, or where the duration is such that full labour and materials fluctuations provisions are required; nor for works of a complex nature or which involve complex services or require more than a short period of time for their execution.'

Not surprisingly, users found this headnote very misleading and it was withdrawn in August 1981 and replaced by Practice Note M2, which is much more indicative of the scope of the contract.

1.2   *The use of MW 80*

Practice Note M2 as revised in September 1988 and March 1994 sets out the criteria for the use of the form. These are:

- Minor building works to be carried out for an agreed lump sum where an architect or contract administrator has been appointed on behalf of the employer. The lump sum price is what the employer will pay, subject of course to the operation of the clauses dealing with variations, fluctuations and the expenditure of provisional sums. MW 80 can be used as a fixed price contract.
- The lump sum offer is based on drawings and/or specifications and/or schedules but without detailed measurements. In principle, the use of bills of quantities is precluded, although it may be that the 'schedules' are in effect scaled down versions of traditional bills.
- The contract period must be such that full labour and materials fluctuation provisions are not required. It is doubtful whether this is a critical factor since, in an appropriate case, a suitable fluctuations clause can be incorporated.
- MW 80 is generally suitable for contracts up to the value of £70,000 at 1987 prices – say £90,000 at 1998 prices. Contract value is not, however, necessarily a deciding factor. A key factor seems to be the complexity of the works.
- Where the employer (through the architect) wishes to control the selection of subcontractors for specialist work the use of the form is prima facie precluded. The practice note points out that this may be done by naming a firm in the tender documents or in a provisional sum instruction, but 'there are no provisions in the Form which deal with the consequences ... nor is there any standard form of sub-contract which would be applicable to such selected sub-contractors'.

Clearly the use of nominated or named subcontractors as such would require substantial amendments to the form as printed. It may also be noted that there is no currently available standard form of subcontract for use with ordinary ('domestic') subcontractors although MW 80 envisages that the contractor may sublet with the architect's consent.

The practice note suggests that the control of selection of specialists can be achieved by the employer contracting directly with the specialist. This may have other unfortunate consequences as noted later. In fact, MW 80 is highly suitable for use on all projects which have a simple work content and where bills of quantities are not required. As with IFC 84 – the next contract up in the JCT range – three things debar the use of MW 80:

(1) Complexity of work
(2) The wish or need for nominated subcontractors and/or nominated suppliers
(3) The wish to make the contractor (or any subcontractor) wholly or partly responsible for design.

MW 80 should not be used merely on account of its apparent simplicity or because, sensibly, the architect dislikes the complex administrative procedures of JCT 80 or the more detailed provisions of IFC 84. The brevity and simplicity of MW 80 is more apparent than real because its operation depends to a large extent on the gaps in it being filled by the general law. Some of the more obvious gaps can be plugged by drafting (or getting an expert in construction contracts to draft) suitable clauses. To take but one example: nowhere in MW 80 is there any provision dealing with contractor's 'direct loss and/or expense' claims as found in JCT 80 and IFC 84, except for the provisions of clause 3.6 which require the valuation of variations to include any direct loss and/or expense incurred by the contractor due to regular progress of the works being affected by compliance with a variation instruction.

This does not mean that the contractor must allow for the possibility of claims in his tender price or that, in appropriate circumstances, he cannot recover them. It merely means that there is no contractual right to reimbursement and that the architect has no power under the contract to deal with them. As explained in Chapter 10, the contractor can pursue such claims in adjudication, arbitration or by means of legal action. In Chapter 10 we also make appropriate suggestions for dealing with this familiar construction industry problem.

1.3   *Arrangement and contents of MW 80*

MW 80 consists of only nine contract conditions, sub-divided in what is not always the most logical way. In addition, there is a separate Supplementary Memorandum. The basic form consists of ten pages. The form is not for use in Scotland which has its own legal system and applicable rules of law. The actual conditions of contract occupy ten pages of the document and are printed in double-column format. The whole document is arranged as follows:

MEMORANDUM OF AGREEMENT
RECITALS
ARTICLES OF AGREEMENT
Table of Contents
CONDITIONS
1 Intentions of the parties
   Contractor's obligations [1.1]
   Architect's/Contract Administrator's duties [1.2]
   Reappointment of planning supervisor or principal contractor –
      notification to Contractor [1.3]
   Alternative B in the 5th recital – notification by contractor –
      regulation 7(5) of the CDM Regulations [1.4]
   Giving or serving of notices or other documents [1.5]
   Reckoning periods of days [1.6]
   Applicable law [1.7]
2 Commencement and completion
   Commencement and completion [2.1]
   Extension of contract period [2.2]
   Damages for non-completion [2.3]
   Completion date [2.4]
   Defects liability [2.5]
3 Control of the works
   Assignment [3.1]
   Subcontracting [3.2]
   Contractor's representative [3.3]
   Exclusion from the works [3.4]
   Architect's/Contract Administrator's instructions [3.5]
   Variations [3.6]
   Provisional sums [3.7]
4 Payment
   Correction of inconsistencies [4.1]
   Progress payments and retention [4.2]

Penultimate certificate [4.3]
Notices of amounts to be paid and deductions [4.4]
Final certificate [4.5]
Contribution, levy and tax changes [4.6]
Fixed price [4.7]
Right of suspension by contractor [4.8]
5 Statutory obligations
Statutory obligations, notices, fees and charges [5.1]
Value added tax [5.2]
Statutory tax deduction scheme [5.3]
[Number not used] [5.4]
Prevention of corruption [5.5]
Employer's obligation – planning supervisor – principal
contractor [5.6]
Duty of principal contractor [5.7]
Successor appointed to the contractor as the principal contractor
[5.8]
Health and safety file [5.9]
6 Injury, damage and insurance
Injury to or death of persons [6.1]
Injury or damage to property [6.2]
Insurance of the works – Fire, etc – New works [6.3A]
– Existing structures [6.3B]
Evidence of insurance [6.4]
7 Determination
Notices [7.1]
Determination by employer [7.2]
Determination by contractor [7.3]
8 Supplementary Memorandum
Meaning of references in 4.5, 5.2 and 5.3
9 Settlement of disputes – Adjudication – Arbitration
Supplementary Memorandum
Part A – Contribution, levy and tax changes
Part B – Value added tax
Part C – Statutory tax deduction scheme
Part D – Adjudication
Part E – Arbitration

There is, happily, little supporting documentation required. It
consists only of a Supplementary Memorandum (see above) and
two practice notes (M1 and M2). Practice Note M1 merely summarises the changes from the earlier edition of the form, while M2
has been summarised in section 1.2.

1.4 Contractual formalities

Like any other contract, a building contract is made by an offer and acceptance and supported by consideration. The contractor's tender is the offer and a properly worded letter of acceptance from the employer (or from the architect on his behalf) will create a binding contract. Acceptance of the contractor's tender must be unqualified. An acceptance subject to further agreement does not result in a contract.

It is sensible for a formal contract to be executed by the parties, and it is best to use the printed form for this purpose. In the private sector the preparation of the contract documentation is the architect's responsibility, and he needs to take great care.

Two copies of the printed form are required, and these must be completed identically, filling in the blanks and making the necessary alterations and deletions. The printed form as completed will then be signed or executed as a deed by the employer and the contractor.

The Articles of Agreement

Page 1 should be completed with the descriptions and addresses of the employer and the contractor. The date will not be inserted until the form is signed or sealed.

The first recital is vital because it is here that a description of 'the Works' must be inserted, and that description is important when considering the question of variations, and for other purposes as well. Only two and a half lines are allowed for this description, so the description must be both precise and succinct. The architect's name (or the name of his firm) must also be filled in. The question of contract documents is discussed in Chapter 3 and the necessary deletions must be made both here and in the second recital.

The second recital deals with pricing and makes provision for the contractor to price the specification, schedules or a schedule of rates, and it should be appropriately completed. If the architect has chosen to go out to tender on the basis of drawings and the contract conditions only and the contractor also submits a schedule of rates, the architect would be very unwise to let the reference to the schedule of rates stand. He did not require it and he should not acknowledge its existence.

The third recital is declaratory, and the contract drawings must be signed by or on behalf of the parties using a simple formula such as: 'This is one of the contract drawings referred to in the Agreement made on [date] between [the employer] and [the contractor]'.

If a quantity surveyor has been appointed, his name or the name of his firm must be inserted in the fourth recital. If not, the recital should be deleted.

The fifth recital was added by Amendment MW9, issued in March 1995. It is in two parts: A and B. Part A deals with the situation where all the Construction (Design and Management) Regulations 1994 apply to the work and the employer has instructed the architect that the design of the works is to comply with the provisions of regulation 13. In this case, part B is to be deleted. Part B is to be used (and part A is to be deleted) if only regulations 7 and 13 apply to the work and the employer has instructed the architect that the design of the works must comply with the provisions of regulation 13.

It is also necessary to complete Articles 2 and 3.

Article 2 sets out the contract sum and this should be written out in full with the figures in brackets. Of course, if fluctuations are applicable, this sum will be subject to the provisions of the supplementary memorandum.

Article 3 defines the term 'architect/the contract administrator' and he must insert his name and address (or that of his firm) in the blank space even though he is also mentioned in the first recital. MW 80 provides for a designated architect and the contract is in fact inoperable without an architect who is charged with the performance of many important duties under the contract terms, as discussed in Chapter 4. Indeed, so important is the architect's role, that the employer *must* appoint a successor architect if the named architect retires, resigns his appointment or dies. Article 3 requires him to do this within 14 days of the original architect's ceasing to act. The successor architect cannot disregard or overrule any certificate or instruction previously given.

In Article 4 all but one of the nominators in Article 4.1 and the appointors in Article 4.2 must be deleted; which are deleted is a matter for the employer, acting on the architect's advice. All deletions or alterations on the recitals, articles of agreement and, indeed, in the contract conditions must be struck through and initialled by the employer and the contractor.

Articles 5.1 and 5.2 were added by Amendment MW9 in March 1995 to deal with the effects of the Construction (Design and Management) Regulations 1995. Article 5.1 is strangely worded. It assumes that the architect will be the planning supervisor. Although it is not immediately obvious, the user of the contract can insert an alternative name. This lack of clarity is worrying, because it is important that the employer understands that the planning supervisor role is not inextricably bound up with the architect's normal duties. Article 5.2 states that the contractor will be the principal contractor. Presumably because the planning supervisor is assumed to be a person and the principal contractor is assumed to be a corporate body, the articles make provision for either ceasing to act in the appointed role, but only the planning supervisor dying. In the event of any of these occurrences the planning supervisor and/ or the principal contractor will be whoever the employer appoints in accordance with regulation 6(5) of the CDM Regulations. If Part A of the fifth recital is deleted, Articles 5.1 and 5.2 must also be deleted.

Attestation

For some reason best known to the Joint Contracts Tribunal, there is only one printed form of attestation, and that provides for the contract to be executed by signature only. However, if it has been decided to execute the contract as a deed it is no longer necessary to seal it. The Companies Act 1989 and the Law of Property (Miscellaneous Provisions) Act 1989 set out the requirements. This is a decision which the employer will have made – on the architect's advice – at pre-tender stage, and there is an important practical difference.

The Limitation Act 1980 specifies a limitation period – the time within which an action may be commenced – of six years where the contract is merely signed by the parties ('a simple contract') or of twelve years where the contract is made as a deed (a specialty contract'). The longer period is beneficial from the employer's point of view and that is why most local and public authorities insist on building contracts being executed as deeds.

Figure 1.1 is the usual form of attestation clause for a deed. Most building contractors operate as limited companies and, by law, they are required to have a company seal which they can use if desired, but it will not alone create a deed. In addition, the document must state on its face that it is a deed. Stamp duty is no longer payable on building contracts executed as deeds unless they are complicated by certain matters such as conveyances or leaseback, which will not normally be the case with MW 80.

Whichever method is used, it is the architect's duty to check over the contract and make sure that all the formalities are in order. All too often these administrative chores are pushed on one side and then when something goes wrong the legal profession has a field day.

Table 1.1 tabulates the decisions which have to be made on the

Figure 1.1
Attestation clause for contract executed as a deed

EXECUTED AS A DEED BY THE EMPLOYER

hereinbefore mentioned namely . . . .

by affixing hereto its common seal
[*if an individual substitute 'whose signature is here subscribed'*]

in the presence of:

OR

acting by two directors/a director and its secretary whose signatures
are here subscribed:

namely . . . .

Signature . . . . Director

and . . . .

Signature . . . . Director/Secretary

AND AS A DEED BY THE CONTRACTOR

hereinbefore mentioned namely . . . .

by affixing hereto its common seal
[*if an individual substitute 'whose signature is here subscribed'*]

in the presence of:

OR

acting by two directors/a director and its secretary whose signatures
are here subscribed:

namely . . . .

Signature . . . . Director

and . . . .

Signature . . . . Director/Secretary

Table 1.1
Filling in the form

Item or clause	Comment
Memorandum	Names and addresses of the parties inserted. Date to be inserted when contract is executed.
Recital 1	The description of the works must be sufficient to identify them clearly. Contract drawing numbers must be filled in. Delete inappropriate documents.
Recital 2	Delete as appropriate.
Recital 3	Make sure contract documents are signed.
Recital 4	QS's name and address to be inserted or deleted.
Recital 5	Delete A or B as appropriate.
Article 2	Contract sum in words and figures inserted.
Article 3	Architect's name and address (or his firm's) inserted.
Article 4.1	Delete 3 of RIBA, RICS, CC, NSCC.
Article 4.2	Delete 2 or RIBA, RICS, CIArb.
Article 5.1	Delete if alternative A of 5th Recital is deleted, otherwise complete details of planning supervisor if not the architect.
Article 5.2	Delete if alternative A of 5th Recital is deleted.
Attestation	Under hand or as a deed. If deed chosen, attestation clause must be provided and the printed version deleted.
Clause 2.1	Dates of commencement and completion to be inserted 'To be agreed' should not be used.
Clause 2.3	Amount of liquidated damages inserted.
Clause 2.5	The usual defects liability is three months. If a different period is required, it should be written in and the 'three months' should be deleted.
Clauses 3.6 and 4.1	The appropriate deletions should be made consistent with the recitals.

Table 1.1 *Contd*	
Item or clause	**Comment**
Clauses 4.2.1/ 4.3	Completed as appropriate to retention percentage chosen. The 5% and $97\frac{1}{2}$% are usual.
Clause 4.5.1.1	If a longer or shorter period is desired, 'three months' should be deleted and another period inserted.
Clause 4.6	This must be deleted if fluctuations do not apply.
Clause 5.2	Delete alternative as required. Clause B1.1 applies only if contractor is satisfied at date of contract that his output tax on supplies to employer are positive or zero tax rate.
Clause 5.5	Unless the employer is a local authority this should be deleted.
Clause 6.2	Insert the amount of insurance cover required.
Clause 6.3A and 6.3B	One or other alternative must be deleted and if 6.3A applies the percentage should be written in.

contract clauses. All necessary decisions as to the applicable clauses and amendments must be made at pre-contract stage, and will have been notified to the contractor accordingly. He is entitled to know the terms on which he is expected to contract. In completing the printed form, the architect must ensure that the final version for execution accords with the terms on which the contractor submitted his tender.

1.5 Problems with the contract documents

Problems can arise with the contract documents which have been prepared for signature or execution as a deed by the architect. The contractor must check them over carefully, checking the printed contract against the information given in the tender documents and noting any discrepancies.

If the contractor discovers mistakes or inconsistencies, he should not execute the documents until the matter is rectified. Figure 1.2 is a suitable pro forma letter for the contractor to send to the architect.

Another problem arises where, as is not uncommon, work starts on site before the contractual formalities are completed. It is bad practice to allow this to happen.

It may be that there is already a binding contract in existence, the parties having agreed on the minimum essential terms and there being an unequivocal acceptance by the employer of the contractor's tender. In such a case, however, it may be unclear whether the arbitration clause is properly incorporated unless expressly stated. Alternatively, there may be no contract at all, and in that event and if no contract came into being, the work done would have to be the subject of a *quantum meruit* ('as much as it is worth') claim and the contractor would be entitled to a fair commercial rate: *Lazerbore* v. *Morrison Biggs Wall* (1993) CILL 896. In general, if a formal contract is executed subsequently, its terms would have retrospective effect: *Trollope & Colls Ltd* v. *Atomic Power Construction Ltd* [1962] 3 All ER 1035.

It is better to avoid the potential difficulties, and a contractor who is asked to start work before the contractual formalities are completed is well advised to write to the architect appropriately: Figure 1.3.

Figure 1.2
Letter from contractor to architect if mistakes in contract documents and no previous acceptance of tender

Dear Sir

PROJECT TITLE

We are in receipt of your letter of the [insert date] with which you enclosed the contract documents for us to sign/execute as a deed [*delete as appropriate*].

There is an error on [describe nature of error and page number of document]. This is not consistent with the tender documents on which our tender is based and we are not prepared to enter into a contract on the basis of the contract documents in their present form.

We therefore return the documents herewith and we look forward to receiving the corrected documents as soon as possible.

Yours faithfully

Figure 1.3
Letter from contractor to architect if contractor asked to commence before contract documents signed

Dear Sir

PROJECT TITLE

Thank you for your letter of the [*insert date*] from which we note that the employer requests us to commence work on site pending completion of the contract documents.

It is our understanding of the situation that we are already in a binding contract with the employer on the basis of our tender of the [*insert date*] and his acceptance of the [*insert date*] on terms incorporated by such tender and letter of acceptance.

If the employer will send us written confirmation that he agrees with our understanding of the situation as expressed in this letter, we will be happy to commence as requested.

Yours faithfully

1.6 Notices, time and the law

Clauses 1.5 and 1.6 have been inserted to give effect to sections 115 and 116 of the Housing Grants, Construction and Regeneration Act 1996. If the method of service of a notice or any other document (i.e., a certificate or an instruction) is not specified in the contract, it may be served by any effective means. That is by any means which has the desired effect. It can be served to any agreed address. If the parties, for one reason or another, cannot agree the address in each case, the second part of clause 1.5 comes into play and service can be achieved by delivery or by pre-paid post to the intended recipient's last known main business address. If the recipient is a corporate body, the address should be its registered or principal office. There appears to be nothing to prevent second class postage being employed although it would not be in the sender's interest to do so.

Where anything is stated to be done within a number of days after a certain date, the counting of days begins immediately after that date. Christmas Day and Good Friday are excluded, together with any day which is a bank holiday under the Banking and Financial Dealings Act 1971.

Clause 1.7 states that the law applicable to the contract is the law of England. That is the case whatever may be the nationality of any of the parties to the contract or anyone connected to it. Even if the works are carried out in Scotland under this contract (unlikely) or in France (even more unlikely) the applicable law will still be the law of England.

CHAPTER TWO
CONTRACT COMPARISONS

2.1 *Introduction*

Advising the employer on which form of contract is best suited to
the particular project is one of the architect's important functions.
This is clear from the schedule of services of the conditions of
engagement (CE/95) where under item E06 it states:

> 'Advise on and recommend form of building contract and explain
> the Client's obligations thereunder.'

In the classic statement of the architect's duties – for breach of which
he may be sued – the sixth duty is put in this way:

> 'To consult with and advise the employer as to the form of con-
> tract to be used (including whether or not to use bills of quan-
> tities) and as to the necessity or otherwise of employing a quantity
> surveyor...'

This sentence – taken from Hudson's *Building Contracts*, 11th edition, page 266, should be engraved on every architect's heart, and
advising the employer on which form of contract to use is one of the
most neglected of his professional functions. It is likely that an
architect would be held professionally negligent if he advised the
use of an unsuitable form of contract and, as a direct result, his client
suffered loss. A substantial body of case law has built up as the
result of using standard contract forms for purposes for which they
were not intended and the architect should certainly explain to his
client the main provisions of whichever form of contract he
recommends.

As regards the JCT contracts, JCT has issued a very generalised
practice note (Practice Note 20), the most useful feature of which is
the comparison which it draws between the range of JCT contracts
currently available. In the private and local authority sectors, the
effective choice for work which has been fully designed by the

employer's consultants is between one of four Joint Contracts Tribunal forms, the form produced by the Association of Consultant Architects (ACA2) and the more recent Engineering and Construction Contract (ECC).

The ACA Form – a second and vastly improved edition of which was published in 1984 and revised from time to time thereafter, most recently in 1998 – is not in our view suitable for truly 'minor works', but – to use JCT terminology – there will be many instances where 'minor works' shade into 'intermediate works'.

The essential point about MW 80 is that the contract conditions provide only a skeleton; but certainly where simplicity of administration is desired, the works are uncomplicated, there is no desire for nominated specialists and bills of quantities are not required, MW 80 should be considered. We recently heard of a contract let on this form where the value was in excess of £3 million. On any view, that must be wrong, not to say negligent, on the part of the architect.

In Chapter 1 of our book *Building Contracts Compared and Tabulated* (2nd edn, 1990), we went into the question of contract choice and comparison in some detail, but for present purposes the critical factors in contract choice are:

- The scope and nature of the work
- The presence or absence of bills of quantities
- Lump sum or approximate price.

In the JCT family – if it is decided that bills of quantities are inappropriate – the effective choice is between the 'without quantities' version of JCT 80, IFC 84 and MW 80. The appendix to JCT Practice Note 20 – the full title of which is 'Deciding on the appropriate form of JCT main contract' – sets out the main differences between these three forms on what it describes as 'matter of major significance in building contracts'. The features which it allocates to MW 80 are:

- *Date of possession*
  MW 80 contains a simple provision (Clause 2.1) stating that the works 'may be commenced' on a specified date. In our view, this particular wording is of no real significance in light of the common law position.
- *Extension of time*
  In contrast to JCT 80 and IFC 84, MW 80 has very brief provisions for extensions of time. The contractor must notify the architect of his need for an extension; and the architect must grant a

reasonable extension of the contract time for reasons beyond the
control of the contractor.

- *Liquidated damages*
  MW 80 has a traditional liquidated damages clause. Deduction
  does not depend on the architect issuing a certificate of non-
  completion, but the employer must give written notice to the
  contractor of intention to deduct no later than the date of issue of
  the final certificate. The clause states simply that 'the contractor
  shall pay ...' the agreed amount.
- *Payment and retention*
  Progress payments to be made by the employer if requested by
  the contractor. The retention is not expressly stated to be trust
  money.
- *Variations and provisional sum work*
  Variations are valued on a 'fair and reasonable basis' – using any
  priced documents.
- *Nominated subcontractors and suppliers*
  MW 80 does not allow for nomination of either – which may well
  be considered to be a plus point! The contract is entirely silent
  about selection of specialists.
- *Fluctuations*
  Fluctuations for contributions, levy and tax fluctuations only.
- *Partial possession*
  There are no provisions for partial possession in MW 80.
- *Contractor's money claims*
  Apart from the provision in clause 3.6 requiring the architect to
  include 'direct loss and/or expense' when valuing a variation
  instruction, there are no provisions for reimbursing the con-
  tractor for 'direct loss and/or expense' – but he can recover under
  the general law.
- *Disputes settlement*
  A simple arbitration agreement. It gives the arbitrator an express
  power to open up, review and revise the architect's decisions.
  Arbitration can be opened up at any time. JCT Amendment
  MW11 has augmented this procedure to include adjudication
  (see Chapter 13).

There are, of course, other significant differences between JCT 80,
IFC 84 and MW 80. The architect cannot assume that he has the
same powers as are conferred on him under the more detailed
forms – because he has not – but this should not deter his recom-
mending MW 80 in those many cases in which it is the most suitable
standard form contract.

2.2 *JCT contracts compared*

Once the decision to use the JCT form of contract for employer-designed works, on a lump sum basis and without bills of quantities, is made, the choice is narrowed to the three best-known and mostly widely used of the JCT forms.

Table 2.1 enables readers to see – it is to be hoped at a glance – the position under MW 80 in comparison with JCT 80 (without quantities) and IFC 84.

Table 2.1
MW 80 clauses compared with those of JCT 80 and IFC 84

MW 80 clause	Description	JCT 80 clause	IFC 84 clause
1	**Intentions of the parties**		1
1.1	Contractor's obligations	2	1.1
–	Quality and quantity of work	14	1.2
4.1	Priority of contract documents	2.2.1	1.3
–	Contract bills and SMM	2.2.2	1.5
–	Custody and copies of contract documents	5.1	1.6
1.2	Further drawings and details	5.4	1.7
–	Limits to use of documents	5.7	1.8
1.2	Issue of certificates by architect	5.8	1.9
–	Unfixed materials or goods: passing of property, etc	16.1	1.10
–	Off-site materials and goods: passing of property, etc	16.2	1.11
1.3	Reappointment of planning supervisor or principal contractor	1.5	1.12
1.4	Notification by contractor – regulation 7(5) of CDM regulations	–	–
1.5	Giving or service of notices	1.5	1.13
1.6	Reckoning periods of days	1.6	1.14
1.7	Applicable law	1.8	1.15
–	Employer's representative	1.7	–
2	**Possession and completion**		2
2.1 2.4	Possession and completion dates	23.1.1	2.1
–	Deferment of possession	23.1.2	2..2

Table 2.1 *Contd*

MW 80 clause	Description	JCT 80 clause	IFC 84 clause
2.2	Extension of time	25	2.3
2.2	Events	25.4	2.4
–	Certificate of non-completion	24.1	2.6
2.3	Liquidated damages for non-completion	24.2	2.7
–	Repayment of liquidated damages	24.2	2.8
2.4	Practical completion	17.1	2.9
2.5	Defects liability	17.2	2.10
3	**Control of the works**		**3**
3.1	Assignment	19.1	3.1
3.2	Subcontracting	19.2	3.2
–	Named persons as subcontractors	–	3.3.1
–	Nominated subcontractors	35	–
–	Nominated suppliers	36	–
3.3	Contractor's person-in-charge	10	3.4
–	Performance specified work	42	–
3.5	Architect's instructions	4.1	3.5
3.6	Variations	13	3.6
3.6 3.7	Valuation of variations and provisional sum work	13.4 13.5	3.7
3.7	Instructions to expand provisional sums	13.3	3.8
–	Quotations	13A	–
–	Levels and setting out	7	3.9
–	Clerk of works	12	3.10
–	Work not forming part of the contract	29	3.11

Table 2.1 *Contd*

MW 80 clause	Description	JCT 80 clause	IFC 84 clause
—	Instructions as to inspection; tests	8.3	3.12
—	Instructions following failure of work, etc	8.4.3 8.4.4	3.13
—	Instructions as to removal of work, etc	8.4	3.14
—	Instructions as to postponement	23.2	3.15
4	**Payment**		**4**
—	Contract sum	14	4.1
4.1	Instructions as to inconsistencies, errors, or omissions	2.3 2.2.2.2	1.4 —
4.2.1	Interim payments	30.1	4.2
4.2.2 and 4.5.2	Interest on late payment	30.1.1.1 and 30.8.5	4.2 and 4.6.2
4.4.2 and 4.5.1.3	Withholding payment	30.1.1.4 and 30.8.3	4.3(c) and 4.6.1.3
—	Advance payment	30.1.1.6	4.2
4.3	Interim payment on practical completion	—	4.3
4.2	Interest in percentage withheld	30.5	4.4
4.4	Computation of adjusted contract sum	30.6	4.5
4.5.1.1	Issue of final certificate	30.8	4.6
—	Effect of final certificate	30.9	4.7
—	Effect of certificates other than final	30.10	4.8
4.6 4.7	Fluctuations	38, 39 and 40	4.9
4.8	Right of suspension by contractor	30.1.4	4.4A

Table 2.1 *Contd*

MW 80 clause	Description	JCT 80 clause	IFC 84 clause
—	Fluctuations; named persons	—	4.10
—	Disturbance of regular progress	26.1	4.11
—	Loss and/or expense matters	26.2	4.12
5	**Statutory obligations, etc**		**5**
5.1	Statutory obligations, notices, fees, and charges	6	5.1
5.1	Notice of divergence from statutory requirements	6	5.2
5.1	Extent of contractor's liability for non-compliance	6	5.3
—	Emergency compliance	6	5.4
5.4	Value added tax	15	5.5
5.3	Statutory tax deduction scheme	31	5.6
5.6	Employer's obligation	6A.1	5.7.1
5.7	Duty of principal contractor	6A.2	5.7.2
5.8	Successor to principal contractor	6A.3	5.7.3
5.9	Health and safety file	6A.4	5.7.4
6	**Injury, damages and insurance**		**6**
6.1 6.2	Injury to persons and property and employer's indemnity	20	6.1
6.1 6.2	Insurance against injury to persons and property	21	6.2
6.3A 6.3B	Insurance in joint names of employer and contractor (new buildings)	22A 22B	6.3A 6.3B
6.3B	Insurance in joint names by employer (existing structure)	22C	6.3C

Table 2.1 *Contd*

MW 80 clause	Description	JCT 80 clause	IFC 84 clause
6.4	Evidence of insurance	22A.2 22B.2 22C.3	6.3A.2 6.3B.2 6.3C.3
–	Insurance for employer's loss of liquidated damages	22D	6.3D
7	**Determination**		**7**
7.1	Determination by employer	27.1	7.1
7.1	Contractor becoming bankrupt, etc	27.2	7.2
–	Corruption – determination by employer	27.3	7.3
7.1	Consequences of determination by employer	27.4	7.4
7.2	Determination by contractor	28.1	7.5
7.2	Employer becoming bankrupt, etc	28.1	7.6
7.2	Consequences of determination by contractor	28.2	7.7
8	**Supplementary memorandum**	–	–
8.1	References to 5th recital	–	–
9	**Settlement of disputes – Arbitration**	**41**	**9**
9.1	Adjudication	41A	9A
9.2	Arbitration	41B	9B
–	Legal proceedings	41C	–
–	Determination by employer or contractor	28A	7.8
–	Consequences of determination by either party	28A.3	7.9
N/A	**Interpretation**		
–	References to clauses, etc	1.1	8.1

Table 2.1 *Contd*

MW 80 clause	Description	JCT 80 clause	IFC 84 clause
—	Articles to be read as a whole	1.2	8.2
—	Definitions	1.3	8.3
—	The architect/supervising officer	Art 3A	8.4
—	Priced specification or priced schedules of work	N/A	8.5

CHAPTER THREE
CONTRACT DOCUMENTS AND INSURANCE

3.1 Contract documents

3.1.1 Types and uses

Within its limitations of use (see Chapter 1), MW 80 can be used with a wide variety of supporting documents. Taken together, they are termed the contract documents. In principle, they may consist of any documents agreed between the parties to give legal effect to their intentions.

A number of options are set out in the first recital and they may be conveniently considered as follows:

- The contract drawings
- The contract drawings and the specifications priced by the contractor
- The contract drawings and schedules priced by the contractor
- The contract drawings, the specification and the schedules, one of which is priced by the contractor.

One of these options, together with the Agreement and Conditions annexed to the Recitals, forms the contract documents. They must *all* be signed by or on behalf of the parties.

Before embarking on a project, the options must be studied carefully to arrive at the combination which is most suitable for the work. Note that the second recital provides that the contractor must price either the specification or the schedules or provide his own schedule of rates. Oddly, it appears that the contractor's own schedule of rates does not become one of the contract documents. (The implications of this are discussed in Chapter 11.)

The contract drawings

On very small works, for which of course MW 80 is very well suited, it may be quite acceptable to go to tender merely on the basis of MW

80 and drawings. This system can be very satisfactory, provided that all the information required by the contractor for pricing purposes is included on the drawings. The architect is not precluded from issuing further information by way of clarification in accordance with clause 1.2 as necessary; indeed it is his duty to do so, but the contractor cannot expect the architect to provide every detail no matter how minute. The contractor is expected to use his own practical experience in constructing the works: *Bowmer and Kirkland Ltd* v. *Wilson Bowden Properties Ltd* (1996) CILL 1157.

Since there is no specification or schedules, the contractor will be expected to provide his own schedule of rates. If this is not intended, the second recital must be deleted in its entirety. The significance of the priced document is that it is to be used to value variations, if relevant, under clause 3.6. Reference to the contractor's own schedule of rates is included in this clause even though it is not one of the contract documents. Some commentators advise that the contractor's own schedule of rates should always be excluded, but it is difficult to see the reason for so doing.

The contract drawings and the specification priced by the contractor

In practice, this is a very common way of dealing with small projects. If the specification is to be priced, great care must be exercised in preparing it. This system involves the contractor in taking off his own quantities and the cost of so doing is likely to be reflected in the tender figure. The organisational expertise which is incorporated into the specification will determine how useful the priced document will be in the valuation of variations.

The contract drawings and schedules priced by the contractor

This is the variant of the last system. Since the works must be specified somewhere, it is likely that the specification element will be incorporated into the schedules. Alternatively, depending on the type of work, it may be feasible to put all the specification notes on the drawings. The schedules will normally be quantified, making this system much easier to price from the contractor's point of view. Indeed, often the schedules are really bills of quantities under another name. The contractor needs to take care, however, that his price is inclusive of everything required to carry out the works. This is true even if some items are missed off the schedules but shown on the contract drawings. The point can be a difficult one. It is discussed in section 3.1.2: clause 4.1 (correction of inconsistencies).

The contract drawings, the specification and the schedules, one of which is priced by the contractor

In practice, this combination would be used on larger works when the schedules would take the form of bills of quantities. The contractor would then normally price the schedules rather than the specification. Once again, the contractor must take care that he prices for everything the contract requires him to do since even if the schedules are in the form of bills of quantities, the employer does not warrant their accuracy, neither does the contract provide that the quantities, if given, take precedence over drawings or specification. Thus, if five doors are listed in the schedules, but it is clear the drawings show seven doors, the contractor must price for seven doors and he will be taken to have done so.

In principle, a schedule of work is always to be preferred over a schedule of rates. The former is capable (or should be capable) of being priced out and added together to arrive at the tender figure. A difficulty may arise because it requires a broad measure of agreement on the method of carrying out the works. The contractor will have difficulty in pricing a schedule of work if he considers that a totally new approach will show greater efficiency. Some two stage method of tendering will probably yield best results in such cases when the contractor can be expected to input his suggestions before the schedules are drawn up. Whether two stage tendering is justified on small projects is another matter.

On the other hand, the figures in a schedule of rates cannot be added together to give the tender figure, and the contractor's own schedule cannot be accepted unless he has justified the calculation of the overall sum from the basis of the schedule. To do otherwise would reduce the valuation of variations to a farce. A schedule of rates is most useful where the total content of the work is not precisely known at the outset. MW 80 can be used in this way, with a little adjustment, but it is better to consider some other forms such as JCT 80 With Approximate Quantities or the Standard Form of Prime Cost Contract (PCC 92).

The numbers of the contract drawings must be inserted in the space provided in the first recital. On large contracts, when bills of quantities are used, it is usual to designate as contract drawings only those small-scale drawings which show the general scope and nature of the work. Under this contract, however, the situation is very different. The total number of drawings is likely to be relatively small and the contractor will need all of them in order to prepare his tender. Since the contractor's basic obligation (clause 1.1) is to 'carry out and complete the Works in accordance with the Contract Documents' the architect must be sure that the contract documents taken together do cover the whole of the work. Therefore, the contract drawings must be:

- The drawings from which the contractor obtained information to submit his tender.
- Sufficiently detailed so that, when taken together with the specification and/or schedules, they include all workmanship and materials required for the project.

The further information which the architect must provide under clause 1.2 may consist of drawings, details and schedules. Provided that they are merely clarifying existing information, there is no financial implication. If, however, they show different or additional or less work or materials than that shown in the contract documents, the contractor will be entitled to a variation on the contract sum. None of the 'further information' constitutes a contract document.

Every contract document must be signed and dated by both parties to avoid any later dispute regarding what is or what is not a contract document. In practice, this means signing every drawing and the cover of the specification and/or schedules. It is suggested that each document is endorsed: 'This is one of the contract documents referred to in the Agreement dated ...' or that some other form of words to the same effect be used.

3.1.2 *Importance and priority*

The contract documents provide the only legal evidence of what the parties intended to be the contract between them. They are, therefore, of vital importance. In the case of dispute, the arbitrator or the court will look at the documents in order to discover what was agreed.

Clause 1.1 lays an obligation upon the contractor to carry out the works in accordance with the contract documents. The question often arises: what is the position if the documents are in conflict? Clause 4.1 states that nothing contained in the contract drawings or the specification or the schedules will override, modify or affect in any way whatsoever the application or interpretation of the printed conditions. If, therefore, the specification were to contain a clause purporting to remove the contractor's entitlement to extension of time due to exceptionally adverse weather, it would be ineffective because of this provision, which has the effect of reversing the normal legal rule of interpretation that type prevails over print. The effectiveness of a clause worded in this way has been upheld in the courts on many occasions: e.g., in *English Industrial Estates Corporation Ltd* v. *George Wimpey & Co Ltd* (1972) 7 BLR 122.

Although the contract effectively sorts out priorities as between the printed form and the other contract documents, it gives no further guidance as far as priorities among the other contract documents are concerned. Clause 4.1 simply states that any inconsistency in or between the contract drawings and the contract specification and the schedules (which, if accepted, will include the contractor's own schedule of rates) must be corrected and if such correction results in addition, omission or other change, it must be treated as a variation under clause 3.6.

The particular circumstances of each case will determine exactly how the inconsistency is to be treated. Two main types of inconsistency are common:

(1)   Where workmanship or materials are covered in one of the contract documents, but omitted from the other
(2)   Where workmanship or materials are shown in one of the contract documents, but omitted from the other.

We will confine our consideration to a contract based on drawings and specification which the contractor has priced. If a schedule is also included, the principle is the same but the facts may be more complex. The first thing to establish is what the contractor has legally contracted to do. Article 1 puts the situation very clearly:

'For the consideration [stated below] the Contractor will in accordance with the Contract Documents carry out and complete the work referred to in the first recital together with any changes made to that work in accordance with this Contract...'

Clause 1.1 also refers to the 'Contract Documents' and that clearly means the documents noted in the first recital, that is, in this case, the contract drawings, the contract specification and the conditions. If, therefore, the inconsistency is, for example, the fact that a handrail is shown without bracket on the drawings, but brackets are specified in the specification, or vice versa, it will be deemed that the contractor has allowed for the brackets in his price. This is probably the case even if the brackets are not in the specification, which the contractor has priced, but are shown on the drawing. In this example, even if the brackets are not shown or mentioned on either document, it is likely that the contractor must supply brackets (presumably the cheapest he can find to do the job) at no additional cost: *Williams* v. *Fitzmaurice* (1858) 3 H&N 844. Much, however, will depend on any general terms which have been included in the specification. An expression such as 'The whole of the materials whether specifically mentioned or otherwise necessary to complete the work must be provided by the contractor' would tend to place the responsibility for omissions squarely on the contractor provided that they could be considered 'necessary' to complete the work. In this case, brackets are obviously necessary to support the handrail.

If the documents are in conflict, the position is rather complicated. Since neither drawings nor specification have priority, it is not clear as to which of them the contractor has had reference in formulating his price. It is tempting to consider that the key document is the specification, since the contractor has priced it. Article 1 and clause 1.1 clearly indicate that this is a wrong view of the situation. In pricing the specification, the contractor must have regard to the totality of the documents. We take the view that, if the documents are in conflict, it is for the architect to instruct the contractor as to which document is to be followed in the particular instance, and the contractor is not to be allowed any addition to the contract sum nor is there to be any omission, the contractor being deemed to have included for whichever option the architect chooses. If, however, the architect solves the problem by omitting the work or changing it to something other than is contained in either of the contract documents, it falls to be valued in accordance with clause 3.6 in the usual way. Although this view may be thought to impose undue hardship on the contractor in certain circumstances, particularly as the inconsistency is due to the architect's oversight, it is the only interpretation that appears to take account of the contract as a whole. The contractor bears a great responsibility to examine the documents thoroughly when pricing. This analysis has an odd result. Although clause 4.1 refers to treatment of the correction as a variation under clause 3.6, it is difficult to envisage any situation in which the contractor would be entitled to have such a variation valued at anything other than a nil amount. If the contract is on the basis of drawings and priced schedules which are in fact fully developed bills of quantities, the situation remains the same. The employer does not warrant the accuracy of the bills in this instance unless a clause is included to that effect. This, of course, is in complete contrast to JCT 80 and points out one of the great dangers in using this form for works of a greater value than that for which it is intended.

3.1.3   Custody and copies

There is no specific provision in the contract regarding the custody of the contract documents and the issue of copies to the contractor. It is probably sensible for the architect to keep the original himself, but if the employer wishes to have it, the architect should be sure to have an exact copy. Although there is no express requirement in the contract, it is good practice for the architect to make a copy for the contractor and certify to him that it is a true copy. This can be simply done by binding all the documents together and inscribing the certificate on each document including the drawings. It is sufficient to state 'I certify that this is a true copy of the contract document dated...'

It must be implied in the contract that the architect provides the contractor with two copies of the contract drawings and specification and/or schedules, otherwise he would be unable to carry out the works. It is usual to provide such copies free of charge.

Any further drawings, details or schedules which the architect is to provide under clause 1.2 are not contract documents. They are intended merely to amplify or clarify the information in the contract documents. He is obliged only to supply such drawings as are 'necessary for the proper carrying out of the Works'. It is an implied term of the contract that he will issue such additional information at the correct time: *R.M. Douglas Construction Ltd* v. *CED Building Service* (1985) 3 Con LR 124 and that the information will be correct: *London Borough of Merton* v. *Stanley Hugh Leach Ltd* (1985) 32 BLR 51. Failure to do so is a breach on the architect's part for which the employer is responsible. The contractor would have a legitimate claim for an extension of time without the necessity for any prior application for the information. If he suffers disruption, he may also have a claim for damages which he could pursue at common law.

3.1.4   Limits to use

The contract contains no express terms to safeguard the architect's interests in the drawings and specification and there is no express prohibition on the employer from using the contractor's rates and prices for purposes other than this contract.

The general law, however, covers the position. The architect retains the copyright in his own drawings and specification (unless he expressly relinquishes it) and neither the contractor nor the employer may make use of them except for the purpose of the project. Strictly, the architect may ask the contractor to return all copies to him after the issue of the final certificate, but in practice it is seldom worth the trouble.

The confidentiality of the contractor's rates and prices is safeguarded by the general rule that a party has a duty not to divulge confidential information to third parties. This is especially true when two parties are bound together in a contractual relationship and to divulge the information would clearly cause harm. The contractor's prices are a measure of his ability to tender competitively and secure work. To divulge his rates to a competitor is a serious matter. In practice, it is not easy for the contractor to ensure, for example, that the quantity surveyor does not make use of his prices to assist him to estimate the cost of other contracts, but that is probably of little consequence.

It may be thought prudent to include a clause in the specification, similar to clause 1.8 of IFC 84, to cover the limitations on the use of documents. It does no harm to remind the parties of their obligations in this respect.

3.2 Insurance

3.2.1 Injury to or death of persons

The insurance and indemnity provisions were amended in 1986 by Amendment MW2 and revised again in March 1994 and August 1996 by Amendment MW10.

Under clause 6.1 the contractor assumes liability for and indemnifies the employer against any expense, liability, loss, claim or proceedings arising out of the carrying out of the works in respect of personal injury or death of any person, except to the extent due to act or neglect of the employer or any person for whom he is responsible. The persons for whom the employer is responsible will include anyone employed by him and paid direct such as a directly employed contractor, the clerk of works (if any) and the architect.

In practical terms, the employer will be responsible for the injury or death of any person only insofar as the injury or death was caused by his or his agent's act or neglect. As a result of the amendment the contractor retains liability even if the employer is to some degree responsible. In effect, in such circumstances, the employer makes a contribution to reflect the extent of his negligence. The employer will then join the contractor as a third party in any action and claim an indemnity from him under this clause. The contractor must take out and maintain and cause any subcontractor to take out and maintain insurances which must comply with all relevant legislation in connection with claims for personal injury or death of any person under a contract of service or apprenticeship with the contractor and which arises out of and in the course of the person's employment. No sum of money is stated, nor is space left for a sum to be inserted. The onus is left on the contractor to insure adequately. This would be cold comfort if the employer found that the contractor was under-insured in the event of a claim and has insufficient funds. We suggest that the architect advises the employer to get the opinion of his insurance broker as to the amount which should be included. Such amount can be stated either on the printed conditions or as a separate clause in the specification. Clause 6.4 gives the employer the right to require evidence that the insurance has been taken out, but there is no provision for the employer to take out the necessary insurance himself and deduct the amount from the contract sum if the contract defaults. Nonetheless, failure on the part of the contractor to insure is a breach of contract for which the employer could recover damages at common law, provided of course he could prove the damage he had suffered. In practice, it is certain that the employer would set off such sums against money payable to the contractor.

The requirement that the contractor insures is stated to be without prejudice to his liability to indemnify the employer. In the case of an insurance company failing to pay in the event of an accident, the contractor is bound to find the money himself. The requirements of clause 6.1 should be satisfied if the contractor and subcontractors already have general insurance cover in appropriate terms in an adequate amount.

It is always wise for the employer to retain the services of an insurance broker to give advice about the insurance provisions of the contract. The broker inspects all the relevant documents and certifies to the architect that they comply with the contract requirements. It is essential that the insurance is in operation from the time the contractor takes possession of the site.

3.2.2 Damage to property

Under clause 6.2, the contractor assumes liability for, and indemnifies the employer against any expense, liability, loss, claim or proceedings arising out of the carrying out of the works in respect of damage to any kind of property other than the works to the extent that it is due to the negligence, breach of statutory duty, omission or default of the contractor or any person engaged by the contractor upon the works or any of their respective employees or agents. The contractor's liability under this clause is limited compared with his liability under clause 6.1. The contractor or subcontractor must be at fault for the indemnity to operate. Amendment MW2 has clarified that the contractor need not be totally at fault for indemnity purposes. If he is partly at fault, the employer has a partial indemnity. The contractor must insure and cause any subcontractor to insure against his liabilities under clause 6.2.

The very instructive case of *National Trust for Places of Historic Interest or Natural Beauty* v. *Haden Young* (1994) 72 BLR 1 provided a clear and sensible explanation of the way this clause should work in connection with clause 6.3B. Unfortunately, the amendments added in 1994 together with Amendment MW10 have changed the wording and hence the meaning. It is now clear that the contractor gives no indemnity under clause 6.2, nor does he insure, in respect of the works, or existing buildings to which the work is being carried out or the contents of such existing buildings. However, if damage is caused to the works through the contractor's fault, the contractor will be liable, because his obligation is to carry out and complete the works in accordance with the contract documents.

The amendments have made the interpretation of this clause extremely complicated and we suggest that each situation be carefully considered. In general the position appears to be that if the existing building and contents are the subject of damage by perils listed in clause 6.3B, they should be covered by the insurance under clause 6.3B even if wholly due to the contractor's negligence. Damage caused by matters other than the perils in clause 6.3B must be rectified at the expense of the contractor if due to his negligence, otherwise it is at the risk of the employer. This interpretation has yet to be tested in the courts.

The contract provides for the employer to stipulate the amount of cover required. The general comments with regard to inspection of documents in section 3.2.1 are also applicable to this clause.

There is no provision in MW 80 for insurance against claims arising due to damage caused by the carrying out of the works when there is no negligence or default by any party. Most standard forms of contract contain such provision which can be invoked if required, to cover specific risks, for example the carrying out of underpinning works to adjoining property. It is not always necessary to take out such insurance and the provision is probably omitted from the contract in view of the minor nature of the works envisaged. The amount of the contract sum, however, is no indication of the possible risk to neighbouring premises and the architect would be prudent to assess each project on its own merits. There is no reason why the employer should not, and every reason why he should, take out the appropriate insurance himself with the advice of his broker if circumstances appear to warrant it.

3.2.3 Insurance of the works against fire etc.

Clause 6.3 deals with the insurance of the works against a collection of specific risks. They are as follows:

'fire, lightning, explosion, storm, tempest, flood, bursting or overflowing of water tanks, apparatus or pipes, earthquake, aircraft and other aerial devices dropped therefrom, riot and civil commotion.'

The clause is divided into two parts, one of which is to be deleted. Each part is applicable to a particular situation as follows:

(1)   6.3A: A new building where the contractor is required to insure
(2)   6.3B: Alterations or extensions to existing structures.

3.2.4 A new building where the contractor is required to insure

The provision is not complex and provides for the contractor to insure in the joint names of the employer and himself against loss or damage caused by any of the specified risks. The insurance must cover the full reinstatement value and include:

- All executed works
- All unfixed materials and goods intended for, delivered to, placed on or adjacent to the works and intended for incorporation
and exclude:

- Temporary buildings, plant, tools and equipment owned or hired by the contractor or his subcontractors.

A percentage is to be added to take account of all professional fees.

Great care must be taken in determining the level of insurance cover. Provision must be made for the possibility that the work is virtually complete at the time of the damage and allowance made for clearing away before rebuilding. Although it is the contractor's responsibility, the architect should satisfy himself that the level of cover is adequate by drafting a letter for the employer to send to the contractor under clause 6.4 and if he is not so satisfied, he must notify the employer and the contractor immediately (Figures 3.1 and 3.2). Despite the absence of express provision, the employer must insure himself if the contractor defaults or fails to insure adequately.

The insurance must be kept in force until the date of issue of the certificate of practical completion. This is the case even if practical completion is delayed beyond the date for completion in the contract. If the contractor already maintains an 'All Risks' policy which provides the same type and degree of cover required under clause 6.3A, it will serve to discharge the contractor's obligations provided it recognises the employer as joint insured and the contractor must produce documentary evidence at the employer's request.

If damage occurs, as soon as the insurers have carried out any inspection they may require, the contractor must 'with due diligence' restore or replace work or materials or goods damaged, dispose of débris before proceeding to carry out and complete the works as before. He is entitled to be paid all the money received from insurance less only the percentage to cover professional fees. It is usual for the money to be released in instalments on certificates, but the contract is silent on the point and it is open to the parties to make any other mutually agreeable arrangement. The contractor is not entitled to any other money in respect of the reinstatement and if there is any element of under insurance or excess, he must bear the difference himself.

3.2.5 Alterations or extensions to existing structures

In the case of an existing building, the employer bears the risk and must insure the works in joint names, the existing structures and

Figure 3.1
Letter from architect to contractor if he fails to maintain an adequate level of insurance under clause 6.3A

Dear Sir

PROJECT TITLE

I refer to my telephone conversation with your Mr [*insert name*] this morning and confirm that your response to the employer's letter of the [*insert date*] shows that you are not maintaining an adequate level of insurance as required by clause 6.3A of the Conditions of Contract.

In view of the importance of the insurance and without prejudice to your liabilities under clause 6.3A I have advised the employer to arrange to take out the appropriate insurance on your behalf. Any sum or sums payable by him in respect of premiums will be deducted from any monies due or to become due to you or will be recovered from you as a debt.

Yours faithfully

Copy: Employer
        Quantity surveyor (if appointed)

Figure 3.2
Letter from architect to employer if contractor fails to maintain an adequate level of insurance under clause 6.3A

Dear Sir

PROJECT TITLE

I am not satisfied that the contractor is maintaining an adequate level of insurance as required by clause 6.3A of the Conditions of Contract.

In view of the importance of the insurance, I advise you to take out the necessary insurance on the contractor's behalf without delay. To this end, I have already advised your broker of the situation and you should contact him immediately. Although the terms of the contract make no express provision for you to act on the contractor's default, it is my opinion that you are entitled to set off the cost of the premium from your next payment to the contractor in this instance.

A copy of my letter to the contractor, dated [*insert date*], is enclosed for your information.

Yours faithfully

contents and all unfixed materials as before. The contractor's temporary buildings etc. are again excluded. There is a proviso that the employer need only insure such contents as are owned by him *and* for which he is responsible. The reason for this is obscure and it takes little imagination to foresee problems arising. There is no requirement that the employer insure for the value of professional fees, but they are his responsibility and if he is wise, he will do so. It would be advisable for the architect to suggest the inclusion of a suitable percentage.

The contractor is entitled to require the employer to produce such evidence as he may need to satisfy himself that the insurance is in force at 'all material times', that is, from the date of possession to the date of practical completion. There is no provision for the contractor to take out insurance himself if the employer defaults, nor to have the amounts of any premiums he pays added to the contract sum. Clearly, however, failure to insure on the part of the employer is a breach of contract, but not one which entitles the contractor to determine his employment in accordance with clause 7.2 (see section 12.3.2) or, it is thought, to treat the contract as repudiated at common law. The architect must not allow the situation to arise and must remind the employer at the appropriate time if insurance is to be his responsibility.

If any loss or damage does occur, the architect is to issue instructions regarding the reinstatement and making good in accordance with clause 3.5. If there is any element of under-insurance, this time it is the employer who must make good the difference. The contractor is entitled to be paid for the work he carried out in the normal way, i.e. following valuation under clause 3.6.

It must be noted that whether the employer or the contractor is responsible and has to insure, *all* damage to the works by fire etc. is removed from the contractor's indemnity under clause 6.2.

Unlike the provisions of similar clauses in JCT 80 and IFC 84, there is no provision in this contract for either party to determine the contractor's employment if it is just and equitable to do so after loss or damage. This is presumably a conscious decision on the part of the JCT, and the position is left to be governed by the general law. The total destruction of the works by fire etc. might bring the contract to an end at common law, but in the majority of cases the parties should be able to come to some mutually acceptable arrangement depending on all the circumstances.

3.3 **Summary**

Contract documents

- They may consist of any documents agreed between the parties
- Taken together, they must cover the whole of the work
- The information provided under clause 1.2 is additional to and not part of the contract documents
- They are the only legal evidence of the contract
- The printed conditions override anything contained in the drawings or specifications
- The correction of inconsistencies may not result in a variation
- The architect has an implied duty to supply the contractor with sufficient copies of the contract drawings and specification and/or schedules for him to carry out and complete the work
- Further correct information necessary must be issued at the correct time or the contractor may claim an extension of time and damages at common law
- No drawing or specification must be used for any purpose other than the contract and the contractor's prices must not be divulged to third parties.

Insurance

- Employer's indemnity covers personal injury and death and damage to property – subject to certain exceptions
- Contractor must insure to cover the indemnities
- The employer has no right to insure if the contractor defaults
- There is no provision for insurance to cover damage to property not the fault of either party
- New work must be insured by the contractor against the specified risks
- Alterations and extensions and existing work must be insured by the employer against the specified risks.

CHAPTER FOUR
ARCHITECT

4.1 Authority and duties

The contract contains express provisions regarding the extent of the architect's authority and the duties imposed upon him. These provisions are discussed in detail below (section 4.2). It would be quite wrong to think, however, that the provisions specifically set out in the contract are the end of the matter. The contract is very brief and, even in the case of a much more comprehensive contract form such as JCT 80, the architect has obligations which are sensible but not always immediately apparent.

From the contractor's point of view, the architect's authority is indeed defined by the contract. Thus if he attempts to exceed this authority, the contractor may, quite rightly, ignore him. In fact, if the contractor carries out an instruction which the contract does not authorise the architect to issue, the employer is under no legal obligation to pay for the results. In such a case the contractor cannot take any legal action against the architect under the contract, to which the architect is not a party, but possibly the architect may find that the contractor has grounds for redress against him direct. Of course the contractor's chances of success will depend on all the circumstances. In view of the fact that the contractor would have no contractual remedy through the main contract, the courts might be prepared to consider an action in tort, but *Pacific Associates Inc* v. *Baxter* (1988) 16 Con LR 90 is an obstacle.

In order to prove negligence, three criteria must be satisfied:

(1) There must be a legal duty of care; *and*
(2) There must be a breach of that duty; *and*
(3) The breach must have caused damage.

To take an example, if the architect instructed the contractor to carry out work on land owned by the employer but outside the site boundaries as shown on the contract drawings, the architect would be acting outside any express or implied authority of the contract.

The contractor would be foolish to carry out such an instruction because he would leave himself open to action by the employer for trespass at the very least. The architect would possibly find himself on the receiving end of actions from both contractor (in tort) and employer (under the architect's terms of engagement).

The architect will, in any event, still be liable to third parties under the rule in *Hedley Byrne & Partners* v. *Heller & Co Ltd* [1963] 2 All ER 575, for negligent statements if it can be shown that the third party acted in reliance on the advice and the architect knew of that reliance and accepted it. He will have liability to his client on this basis which will run together with his liability under the conditions of his engagement: *Richard Roberts Holdings Ltd* v. *Douglas Smith Stimson & Partners* (1988) CILL 432. Indeed, an architect's tortious liabilities may exceed those contained in the conditions of engagement; *Holt* v. *Payne Skillington* (1995) TLR 18 December 1995. The courts have recently shown signs of moving the boundaries of *Hedley Byrne* liability to include duties other than statements: *Henderson* v. *Merrett Syndicates* (1994) 69 BLR 29; *Conway* v. *Crowe Kelsey and Partners* (1994) CILL 927. Actions by contractors against architects based on reliance on negligent drawings, specifications and possibly contract administration are becoming a possibility again.

The architect's powers and duties (see Tables 4.1 and 4.2) flow directly from his agreement with the employer. It is always prudent for them to enter into a formal written contract following the terms of the RIBA Standard Form of Agreement for the Appointment of an Architect (SFA 92) or Conditions of Engagement (CE/95) which is probably more appropriate for small works. The actual small works version (SW/96) may be too simple for all but the smallest works. This contract will determine the precise extent of the architect's authority, but it is important to remember that, whatever the terms of his appointment may be, it has no effect on the building contract between contractor and employer. Therefore, provided that the exercise of the architect's authority is within the limits laid down in the building contract, the contractor may safely carry out his instructions, take notice of his certificates, etc. without worrying whether the architect has obtained the employer's consent.

For example, the Conditions of Appointment state (clause 3.1.4) that the architect must not make any material amendment to the approved design without the employer's consent, and lay an obligation upon the architect to notify the employer if the services, the fees, any other part of the appointment or any information or approvals need to be varied (clause 1.2.7). The contract, however, empowers the architect (clause 3.5) to issue instructions which may

Table 4.1
Architect's powers under MW 80

Clause	Power	Comment
2.5	Instruct contractor not to make good defects	No power to deduct money equal to the cost of making good
3.2	Consent in writing to subcontracting	Consent must not be unreasonably withheld
3.4	Issue instructions requiring the exclusion from the works of any person employed thereon	The power must not be exercised unreasonably or vexatiously
3.5	Issue written instructions	
3.5	Require the contractor to comply with an instruction by serving written notice on him	The contractor has 7 days from receipt of the written notice in which to comply. If he fails to do so the employer may employ and pay others to carry out the work and may recover the costs involved from the contractor
3.6	Order an addition or to omission from or other change in the works or the order or period in which they are to be carried out	The variation is to be valued by the architect on a fair and reasonable basis unless its value is agreed with the contractor before the instruction is carried out
3.6	Agree the price of variations with the contractor prior to the contractor carrying out the instruction	
7.2.1	Give notice to the contractor specifying a default and requiring it to be ended	The contractor has 7 days in which to end the default

Table 4.2
Architect's duties under MW 80

Clause	Duty	Comment
1.2	Issue any further information necessary for the proper carrying out of the works	
1.2	Issue all certificates	See Table 4.3
1.2	Confirm all instructions in writing	Clause 3.5 specifies the procedures
2.2	Make in writing such extension of time as may be reasonable	If it becomes apparent that the works will not be completed by the stated completion date and the causes of delay are beyond the control of the contractor including compliance with an architect's instruction not due to the contractor's default and the contractor has so notified the architect
2.4	Certify the date when in his opinion the works have reached practical completion and the contractor has complied sufficiently with clause 5.9	
2.5	Certify the date when the contractor has discharged his obligations in respect of defects liability	The contractor's obligation is limited to remedying defects, excessive shrinkages or other faults appearing within the defects liability period and which are due to materials or workmanship not in accordance with the contract *or* to frost occurring before practical completion
3.5	Confirm oral instructions in writing	This must be done within 2 days of *issue*

Table 4.2 *Contd*		
Clause	**Duty**	**Comment**
3.6	Value variation instructions	The value is to be on a fair and reasonable basis and where relevant the prices in the priced documents must be used. Any direct loss and or expense incurred by the contractor must be included if due to regular progress affected by compliance with instructions or due to compliance or non-compliance by the employer with clause 5.6
3.7	Issue instructions as to expenditure of provisional sums	The instruction is to be valued as a variation under clause 3.6
4.1	Correct inconsistencies in or between the contract drawings, specification and schedules	If the correction results in a change it is to be valued as a variation under clause 3.6
4.2.1	Certify progress payments to the contractor	Such payments are to be certified at intervals of not less than 4 weeks calculated from commencement
4.3	Certify payment to the contractor of 97.5% of the total amount to be paid to him	This must be done within 14 days of the date of practical completion certified under clause 2.4
4.5.1.1	Issue a final certificate	Provided the contractor has supplied all documentation reasonably required for the computation of the final sum and the architect has issued his clause 2.5 certificate (defects liability). The final certificate must be issued within 28 days of receipt of the contractor's documentation
6.3A	Issue certificates	To pay insurance money to the contractor
6.3B	Issue instructions for the reinstatement and making good of loss or damage	Should loss or damage be caused by clause 6.3B events

both alter the design and increase the total cost of the works. The contractor must carry out such instructions and the employer is bound to pay. If the employer did not consent to the alteration in design or the increased expenditure, his remedy lies against the architect.

If the architect fails to carry out his duties under the contract, it will generally be held to be a default for which the employer will be liable to the contractor: *Croudace Ltd* v. *London Borough of Lambeth* (1986) 6 Con LR 72. Failure to certify progress payments at the right time is an example of such a default, but not failure to include the amount the contractor considers is due: *Killby & Gayford* v. *Selincourt Ltd* (1973) 3 BLR 104.

In performing his duties the architect will be expected to act with the same degree of skill and care as the average competent architect: see *Chapman* v. *Walton* [1833] All ER Rep 826. If he professes to have skills or experience which are greater than average, he may be judged accordingly. Thus, if he holds himself out as, for example, an expert in historic buildings, he will be expected to show a higher degree of skill in that particular area. If something goes wrong, to say that fellow architects without that specialisation would have taken the same action will be insufficient defence. It would have to be shown that a body of specialised opinion would have come to the same conclusion. The body need not be substantial in numbers provided that it exists and is responsible: *De Freitas* v. *O'Brien* [1995] PIQR P281.

This is not the place to discuss in detail the architect's general duties to his client and to third parties. One aspect of those duties, however, does affect his administration of the contract. That is his duty to his client to be familiar with those aspects of the law which affect his work. He is not expected to have the detailed knowledge of a specialist building contracts lawyer, but he must be capable of advising his client regarding the most suitable contract for a particular project. This implies that he must have, at least, a working knowledge of the main forms of contract. In addition, he must be able to give detailed advice on the contract of his choice. Although problems may arise during the currency of the contract which clearly demand the attentions of a legal expert, the architect's client will be less than impressed and may well consider him to be incompetent if he is unable to explain the basic provisions of the contact and deal with the day-to-day running of the job without legal assistance. If his client is put to unnecessary expense due to the fact that the architect had inadequate knowledge of the contractual provisions, the client may well sue. The architect is also expected to be aware of decisions of the courts relevant to his field of knowledge.

Up to the moment the employer and contractor enter into the contract, the architect has been acting as agent with limited authority for the employer. During the contract he is expected to continue to act as the employer's agent, but also to administer the terms of the contract fairly between the parties. The employer may find the change difficult to understand and, to avoid problems, the wise architect will explain the situation to him, perhaps at the same time as he sends the contract documents for signature Figure 4.1. The architect is not in a quasi-arbitral role and he is not immune from any action for negligence by either party. The point was established by *Sutcliffe* v. *Thackrah* [1974] 1 All ER 319.

The result is that the architect may find himself in the position of deciding whether he is in default himself, e.g. in considering an extension of time. In the case of *Michael Sallis & Co Ltd* v. *Calil and W.F. Newman & Associates* (1987) 12 Con LR 68, it was held that an architect owes a duty of care to a contractor when the contract calls on the architect to act fairly between the parties. Such things as extensions of time and certification of money would fall under this heading. The decision was called into question by the Court of Appeal in *Pacific Associates Inc* v *Baxter* (1988) 16 Con LR 90, a case which concerned a firm of consulting engineers. The court held that the contractor had a contractual route, through arbitration, to obtain satisfaction, there was no reliance on the engineer, and the contract expressly excluded the engineer's liability. The latter case is currently the law, but since the route to the decision was somewhat different in each case, the arguments put forward in *Sallis* may yet be revived in a higher court particularly in the light of the *Henderson* and *Conway* decisions noted above.

If the employer suffers loss or damage due to the fact that the architect has issued information late and the architect has granted an extension on that basis, he may well seek to recover that loss from the architect. It might be thought that the architect's position is slightly less hazardous under this contract than it is under JCT 80 or IFC 84 because of the absence of a full provision enabling the contractor to claim 'loss' and 'expense' under the contract as a result of architect's defaults, e.g. through late supply of information. However, if the contractor brought a successful action against the employer at common law, the employer might well be able to recoup his loss from the architect. Even if no such action was brought, the employer may sue the architect for loss of liquidated damages.

Figure 4.1
Letter from architect to employer explaining his duty to act impartially

---

Dear Sir

PROJECT TITLE

[*Insert the main point of the letter and then continue:*]

This is probably an opportune moment to explain the nature of the additional responsibility which I carry during the currency of the contract. Until the contract is signed I am required, in accordance with the conditions of my appointment, to act solely as your agent within the limits laid down in those conditions. Thereafter, although I continue to act for you as before, I have the additional duty of administering the contract conditions fairly between the parties. In effect, this means that I must make any decisions under the contract strictly in accordance with the terms of the contract. If you require any further explanation of the position, I would be delighted to meet you for that purpose.

Yours faithfully

The extent of the authority and duties of an employee of the employer, for example in local or central government, is not always very clear. If the architect is in this position, the contractor may well consider that he is acting as agent for the employer and that any instruction which he may issue which is not expressly empowered by the contract is, in effect, a direct instruction from the employer. The dangers of incurring unexpected costs are obvious and the architect should ensure that the extent of his authority is made clear to the contractor at the beginning of the contract (Figure 4.2).

As architect employee, his duty to act fairly between the parties remains, but it is admittedly difficult for the architect to convince the contractor that he is acting impartially. The administration of the contract under these circumstances calls for complete integrity, not only on his part but also on the part of the employer who has a duty to ensure that the architect carries out his duties properly in accordance with the contract: see *Perini Corporation* v. *Commonwealth of Australia* (1969) 12 BLR 82. Although it is usual for the actions of local authority employees to be governed by Standing Orders of the Council, they are of no concern to the contractor unless they have been specifically drawn to his attention at the time of tender. Thus an order that only chief officers may sign financial certificates would not avail the architect as a defence if he was unable to issue a certificate because the chief officer was unavailable. The contractor is entitled to rely on the architect's apparent authority. Orders that certificates may not be issued without authority from the audit department are likewise irrelevant so far as the contractor is concerned.

The best summary of the legal situation is this:

'An architect is usually and for the most part a specialist exercising his special skills independently of his employer. If he is in breach of his professional duties he may be sued personally. There may, however, be instances where the exercise of his professional duties is sufficiently linked to the conduct and attitude of the employers so as to make them liable for his default.' (Kilner Brown J in *Rees & Kirby Ltd* v. *Swansea Corporation* (1983) 25 BLR 129)

Although this case subsequently went on appeal, it is suggested that Kilner Brown J's statement remains an accurate expression of the law.

Figure 4.2
Letter from architect to contractor if architect is employee in local authority, etc.

Dear Sir

PROJECT TITLE

Possession of the site will be given to you by the employer on the [*insert date*] in accordance with the contract provisions.

In order to avoid any misunderstandings which might arise in the future I should make clear, as far as you are concerned, that the extent of my authority is laid down in the contract. Although I am an employee of [*insert name of employing body*], I have no general power of agency to bind the employer outside the express contract terms. If I have cause to write to you on behalf of the employer, I will clearly so state. If any matters fall to be decided by me under the contract, I will make such decisions impartially between the parties.

Yours faithfully

4.2 Express provisions of the contract

Article 3 of MW 80 provides for the insertion of the name of the architect. The person so named will be the person referred to in the conditions whenever the word 'architect' appears. It is common practice to insert the name of the architectural firm (or the name of the chief architect in a local authority) because of the difficulties which could arise in having to renominate every time the project architect left the firm, died or retired. It is seldom the cause of any dispute, but it is prudent to notify all interested parties of the name of the authorised representative, i.e. the project architect, who will administer the contract on a day-to-day basis (Figure 4.3). Changes in the identity of the project architect should be notified in the same way. The case of *Croudace Ltd* v. *London Borough of Lambeth* (1986) 6 Con LR 72, establishes that at common law the employer is under a duty to appoint another architect if, for example, the named architect resigns, retires or dies, and Article 3 make this duty explicit.

It is not strictly necessary for the architect named in Article 3 to sign all letters, notices, certificates and instructions personally, but they must be signed by a person duly authorised. It is not wise for the authorised representative to sign his name only, even though he may be using headed stationery, because:

- The letter, certificate, etc. must be signed by or in the name of the architect named in the contract: *London County Council* v. *Vitamins Ltd* [1955] 2 All ER 229.
- Otherwise the letter may be deemed to be written merely on the authority of the person who signs, with serious financial consequences.

The architect should not:

- Use a rubber stamp. It is not good practice although it is probably valid.
- Sign someone else's name and add his initials. To sign in someone's name is best accomplished by procuration ('per pro' or 'pp').

If the named architect dies or ceases to act for some other reason, it is the employer's duty to nominate a successor within 14 days of the death or ceasing to act of the named architect: *Croudace Ltd* v. *London Borough of Lambeth* (1986) 6 Con LR 72. Unlike the position under JCT

Figure 4.3
Letter from architect to contractor naming authorised representatives

Dear Sir

PROJECT TITLE

This is to inform you formally that the architect's authorised representatives for all the purposes of the contract are:

[*insert name*] – Partner in charge of the contract.
[*insert name*] – Project architect.

Until further notice the above are the only people authorised to exercise the authority of the architect in connection with the contract.

Yours faithfully

for and on behalf of [*insert the name of the architect/practice in the contract*]

Copies: Employer
          Quantity surveyor
          Consultants
          Clerk of works

80, there is no provision for the contractor to object to the successor, but there is nothing to prevent the contractor from referring the matter to immediate arbitration if he feels strongly about it. The contractor must, however, continue with the works while awaiting the outcome of the arbitration and, in the majority of contracts carried out under this form, it is likely that the work will be complete by the time a decision is reached. That is probably the reason for the exclusion of the right of objection in the first place. Clearly, the employer would do well to listen carefully if the contractor makes any representations, in order to avoid such difficulties.

There is an important proviso to this Article which states that 'no person subsequently appointed to be the Architect ... under this Contract shall be entitled to disregard or overrule any certificate or instruction given by' the former architect. This is a sensible provision to safeguard the contractor's interests under circumstances which are in the sole control of the employer. A successor architect who disagrees with some of his predecessor's decisions should notify the employer immediately (Figure 4.4), to safeguard his own position. The proviso clearly cannot mean that certificates or instructions given by the former architect must stand for all time, even if they are wrong. The implication of the proviso is that if it is necessary for the successor architect to make changes, they will be treated as variations and the contractor will be entitled to payment accordingly. For example, if the former architect had given instructions for the construction of a detail which, in the opinion of the new architect, would lead to trouble, the new architect could issue further instructions correcting the matter and the contractor would be paid for correcting the work. Unlike instructions, certificates are expressions of the architect's opinion in tangible form for the purposes specified in the contract: *Token Construction Co Ltd* v. *Charlton Estates Ltd* (1973) 1 BLR 48. Certificates of practical completion and making good defects are not, in any event, susceptible to change. On the other hand, an architect is normally entitled to amend a financial certificate at the next time for payment and, being cumulative, it is difficult to see how a new architect could forfeit this right.

Although clause 1.1 is titled 'Contractor's obligation,' it creates a problem for the architect. After concisely stating the contractor's obligation to carry out and complete the works with due diligence and in a good and workmanlike manner in accordance with the contract documents, there is a proviso:

'where and to the extent that approval of the quality of materials or of the standards of workmanship is a matter for the opinion of

Figure 4.4
Letter from architect to employer if disagreeing
with former architect's decisions, etc.

Dear Sir

PROJECT TITLE

I have now had the opportunity of examining all the drawings, files and other papers relating to this contract and I have visited the site and spoken to the contractor.

Article 3 of the contract prevents me from disregarding or overruling any certificate or instruction given by the architect previously engaged to administer this contract. This does not prevent me from issuing further certificates and, particularly, further instructions which may vary instructions given by the former architect. Of course in the latter instance there will be a cost implication.

I list below the matters on which I find myself unable to agree entirely with the decisions of the former architect:

[*list all matters which which you disagree*]

When you have had the opportunity to study these matters I suggest we should meet to discuss ways of dealing with them.

Yours faithfully

the architect ... such quality and standards shall be to the reasonable satisfaction of the architect...'

It has been held that such a proviso in the JCT 80 and IFC 84 forms of contract must be given a very wide meaning: *Crown Estates Commissioners* v. *John Mowlem & Co Ltd* (1994) 70 BLR 1; *Colbart Ltd* v. *H. Kumar* (1992) 59 BLR 89. In effect, the quality and standards of all materials and workmanship are matters for the architect's reasonable satisfaction. If the contractor disagrees that the architect is being reasonable, he can, of course, seek adjudication on the point. The danger is threefold:

- It might be argued that it is implicit in the proviso that the architect will in due course and especially if requested by the contractor, express his satisfaction about all matters. The contract provides no machinery for so doing, but the wise contractor will ask the architect to express his satisfaction in writing (see Figure 4.5) and it is difficult to see how the architect could be justified in refusing unless he does not approve.
- The final certificate is not stated to be conclusive about anything, thus avoiding the trap which used to exist in JCT 80 and IFC 84, but it would not be easy for the architect to justify issuing the final certificate and making himself *functus officio* if there are matters with which he is dissatisfied.
- If the architect formally approves quality and standards, his approval will override any requirements in the contract documents, so if he approves something which is not in accordance with the contract, the employer is prevented from seeking redress from the contractor on the ground of lack of compliance.

Clause 1.2 very briefly sets out the architect's duties. These are:

- The issue of further information
- The issue of all certificates in accordance with the contract (see Table 4.3)
- The confirmation of all instructions in writing in accordance with the contract.

The amount, if any, of further information necessary will vary greatly depending on the size and complexity of the work. In the case of small simple projects, the contract drawings may need very little elaboration. Whether the information is necessary should be a

Figure 4.5
Letter from contractor to architect seeking approval

Dear Sir

PROJECT TITLE

We refer to item [*insert item number*] which states that [*specify material or work*] is to be to your approval. Clause 1.1 of the conditions of contract state that in such instances the quality and standards are to be to your reasonable satisfaction.

The above mentioned material has been installed/work has been completed [*delete as appropriate*] and we should be pleased to receive your written approval so as to comply with the terms of the contract.

Yours faithfully

Table 4.3
Certificates to be issued by the architect under MW 80

Clause	Certificate
2.4	Practical completion
2.5	Making good of defects
4.2.1	Progress payments
4.3	Penultimate certificate
4.5.1.1	Final certificate
6.3A	Insurance payments

matter of fact, not opinion. The contractor is expected to use his own practical experience in carrying out the work: *Bowmer & Kirkland Ltd v. Wilson Bowden Properties Ltd* (1996) CILL 1157. Note that there is no obligation on the contractor to apply for further information either in this clause or anywhere else in the contract. The prudent contractor will do so, but it is the architect's duty to supply correct information to enable the contractor to carry out the works properly in accordance with the contract: *London Borough of Merton* v. *Stanley Hugh Leach Ltd* (1985) 32 BLR 51. The completion date is part of the provisions of the contract, therefore the architect's duty is to supply necessary information at the correct times to enable the contractor to proceed with the works and to complete them by the contract completion date. If he fails to do so, it will be a breach for which the contractor can claim extension of time and possibly damages at common law.

The issue of certificates refers not only to progress payments but also to such things as practical completion and making good of defects after the defects liability period. Although there is no pre-scribed way in which a certificate should be set out, it is a formal document. It may be in the form of a letter, but for the avoidance of doubt it is always good practice to head the letter 'Certificate of ...' and begin 'I certify ...'. Where a number of certificates are to be issued in sequence it is normal to number them in order. Delay in the issue of a certificate is a serious matter and may give rise to financial claims by the contractor (Table 10.1).

Confirming instructions will be dealt with in section 4.3.

The remaining express duties of the architect are covered in the appropriate chapters of this book. Because this contract is so brief, his duties are also briefly stated. This can be misleading and the architect should always bear in mind that the courts will imply terms to cover the way in which he must administer the contract. Generally, he is expected to act promptly and efficiently. If he does, he will avoid claims and contribute towards the smooth running of the contract.

4.3   Architect's instructions

All instructions to the contractor must be confirmed in writing to be effective. Despite what some contractors maintain, there is abso-lutely no requirement that instructions must be on a specially printed form headed 'Architect's Instruction', although it is undoubtedly good practice to use such forms because:

- It leaves no room for doubt that the architect is issuing an instruction. An instruction buried in a letter which deals with a great many other things can sometimes be confusing.
- It makes the job of keeping track of instructions for the purposes of valuation and checking so much easier.

Instructions can be given in letter form, provided it is made clear that the architect is issuing an instruction (Figure 4.6). He can also give a hand-written instruction on site, provided it is signed and dated (the architect should always sign on behalf or in the name of the architect named in the Articles if it is someone other than himself). Instructions contained in the minutes of site meetings are valid if the architect is the author of the minutes and if they are recorded as agreed at a subsequent meeting. It is probable, however, that such instructions are not effective until the contractor receives a copy of the minutes recording agreement. Since site meetings are sometimes at monthly intervals, the matter giving rise to the instruction may be ancient history before the contractor's duty to comply becomes operative; minuted instructions are, therefore, best avoided.

The position with regard to drawings is uncertain. If the architect issues a drawing together with a letter instructing the contractor to use the drawing for the works, that is certainly an instruction for contract purposes. If he simply issues a drawing under cover of a compliments slip, the drawing may be an instruction or it may simply be sent for comment; in most cases it will be taken to be an instruction. The contractor would be wise to check with him first, but the architect may have difficulty showing that the drawing was not intended as an instruction if the contractor simply carries out the work. Note that if the architect ends the contractor a copy of the employer's letter requiring some action under cover of a compliments slip, it may not be an instruction at all. The moral is clear: compliments slips should not be used where an instruction is intended.

Clause 3.5 is the principal contract provision governing the issue of instructions. The procedure is shown in Figure 4.7. Despite what, at first sight, may appear to be an all-embracing provision – 'The Architect ... may issue written instructions which the Contractor shall forthwith carry out' – the architect must act within the scope of authority and the instruction must relate to the contract works. The courts do not generally favour sweeping generalisations and prefer to see respective rights and duties expressed in precise terms. In the event of a dispute, therefore, it is probable that the clause will be

Figure 4.6
Letter from architect to contractor issuing an instruction

Dear Sir

PROJECT TITLE

In accordance with clause 2.5/3.4/3.5/3.6/3.7/6.3B [*delete as appropriate*] please carry out the following instructions forthwith:

[*insert the instruction*]

Yours faithfully

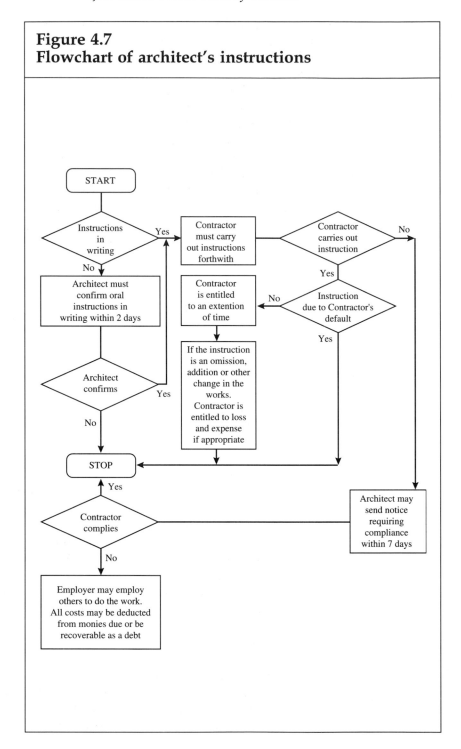

Figure 4.7
Flowchart of architect's instructions

given a very narrow interpretation. The contractor must carry out the instruction as soon as he reasonably can: *London Borough of Hillingdon* v. *Cutler* [1967] 2 All ER 361.

Provision is made for the situation if instructions are issued orally. They must be confirmed in writing by the architect within two days. The contract is silent as to when such an instruction becomes effective. It is considered that, in context, the instruction would not become effective until the contractor receives the confirmation. Although the practice of issuing oral instructions is widespread, it is difficult to see why it was necessary to make special provision in the contract since the effect of the confirmation is no different from the effect of a simple written instruction in the first place. The oral instruction itself is irrelevant. Note that there is no provision for the contractor to confirm oral instructions. If the contractor confirms an oral instruction, the architect remains silent and the contractor proceeds with the work, it is likely that the employer will be estopped (prevented) from denying the contractor's entitlement to payment: *Bowmer & Kirkland Ltd* v. *Wilson Bowden Properties Ltd* (1996) CILL 1157. Ideally, oral instructions should be avoided.

If the contractor does not carry out instructions 'forthwith', the contract provides a remedy in much the same form as JCT 80, clause 4.1.2. As a first step, the architect must send the contractor a written notice requiring him to comply with the instruction within 7 days (Figure 4.8). If the contractor does not so comply, the architect should advise the employer (Figure 4.9) that he may employ others to carry out the work detailed in the instruction. Although the contract specifically states that it is the employer who may employ others, he will expect the architect to advise him and handle the details. It will amount to a completely separate contract and, in order to ensure that there can be no reasonable question about costs, the employer is entitled and the architect should obtain competitive quotations from three firms if time and circumstances permit: *Rhuddlan Borough Council* v. *Fairclough Building Ltd* (1985) 3 Con LR 38. After the work is completed, the employer has the right to deduct all costs from monies due or to become due to the contractor under the contract.

Alternatively, but less attractively for the employer, the monies may be recovered as a debt. In some instances this may be the only practicable way of obtaining the cash, but is best avoided if possible because it involves the always uncertain and costly process of arbitration. Note that the 'costs incurred thereby' refer to additional costs, over and above the cost of the instruction. It is not intended

Figure 4.8
Letter from architect to contractor requiring compliance with instructions before default action taken

RECORDED DELIVERY*

Dear Sir

PROJECT TITLE

On [*insert date*] I instructed you to [*specify*] under clause 3.5 of the contract. That clause requires you to carry out my instructions forthwith.

You have failed to so comply and in accordance with clause 3.5, I hereby require you to carry out the above-mentioned instruction. Should you fail to comply within 7 days after receipt of this notice, the employer will engage others to carry out the work and deduct all the resultant costs from monies due to you under the contract.

Yours faithfully

Copies: Employer
             Quantity surveyor (if appointed)

* Service by recorded delivery is not required by clause 3.5 but is a desirable precaution.

Figure 4.9
Letter from architect to employer if contractor fails to comply with instruction within 7 days after notice

Dear Sir

PROJECT TITLE

I refer to my letter to [*insert name of contractor*], dated [*insert date*], requiring him to comply with my instruction of the [*insert date*] within 7 days.

The 7 days expired yesterday and, during a site visit this morning, I observed that the contractor had not complied with my instruction.

I advise that you should take advantage of the remedy afforded by clause 3.5 of the Conditions of Contract and, if you will let me have your written instructions to that effect, I will obtain competitive tenders from other firms for the carrying out of the work. All additional costs, including my additional fees, can be deducted from future payments due to the contractor and I will liaise with you regarding these details at the appropriate time.

Yours faithfully

that the employer should get the work done at the contractor's expense.

The additional costs will include such items as scaffolding, cutting out and making good, depending on circumstances. The architect is entitled to charge additional fees and the employer may include them in his computation of costs together with any other incidental expenses attributable to the contractor's non-compliance. When deducting the amount, the contractor is entitled to a brief statement showing how the figure has been calculated. Although the wording in JCT 80 is slightly different, it is not thought that it makes any practical difference.

Note that there is no provision for the contractor to object to any instruction, or request the architect to state the empowering provision. However, Figure 4.10 is the kind of letter the contractor may send when receiving a compliance notice. Apart from clause 3.5, there are only five other clauses empowering the architect to issue instructions in particular circumstances:

(1) Clause 2.5 empowers the architect to instruct the contractor not to make good defects at his own cost (this provision is discussed in detail in Chapter 9).

(2) Clause 3.4 empowers the architect to issue instructions requiring the exclusion from the works of any person employed thereon. The instruction is not to be issued unreasonably or vexatiously and is clearly intended to enable the architect to have incompetent operatives removed.

(3) Clause 3.6 empowers the ordering of variations by way of additions, omissions or other changes in the works or the order or period in which they are to be carried out.

(4) Clause 3.7 requires the architect to issue instructions regarding the expenditure of any provisional sum (see Chapter 11).

(5) Clause 6.3B requires the architect to issue instructions for the reinstatement and making good of loss or damage caused by fire, lighting, storm, etc. (see Chapter 3).

Clause 3.3 provides that any instructions given to the contractor's representative on the works are deemed to have been issued to the contractor. In the context of the contract as a whole, such instructions must be in writing for the clause to have any effect, although in practice the architect seldom gives written instructions to the person in charge. Site instructions tend to be oral and are therefore ineffective, written instructions being sent to the contractor's main office.

Figure 4.10
Letter from contractor to architect on receipt of 7 day notice requiring compliance with instruction

---

Dear Sir

PROJECT TITLE

We have today received your notice dated [*insert date*] which you purport to issue under clause 3.5 of the above contract.

[*Add either:*]

We will comply with your instruction Number [*insert number*] dated [*insert date*] forthwith, but our compliance is without prejudice to and reserving any other rights or remedies which we may possess.

[*Or:*]

We consider that we have already complied with your instruction Number [*insert number*] dated [*insert date*]. If the employer attempts to employ other persons and/or deduct any monies due or to become due to us under the contract this will constitute a serious breach of contract for which we shall seek appropriate remedies. However, without prejudice to the foregoing, provided you immediately withdraw your purported notice of non-compliance, our Mr [*X*] will be happy to meet you on site or at your office to sort out what appears to be an unfortunate misunderstanding.

[*Or:*]

It is not reasonably practicable for us to comply within the period specified by you because [*give reasons*]. You may be assured that we are well aware of our contractual obligations and intend to carry out your instruction Number [*insert number*] dated as soon as [*indicate operation, etc*] is complete [*or as appropriate*]. In light of this explanation, perhaps you will be good enough immediately to withdraw your notice requiring compliance.

Yours faithfully

Copy: Employer

It has already been mentioned that the courts will probably construe clause 3.5 narrowly as regards the architect's powers to issue instructions. Although not expressly stated, it is considered that the following common situations will fall within his powers under the clause:

- The correction of inconsistencies between the contract documents. Although clause 4.1 deals with this matter, it does not specifically provide that the architect can issue instructions. Nonetheless, it is considered that he certainly has that power.
- Opening up of work for inspection and testing of materials. If the work or materials are found to be in accordance with the contract, the contractor will have a claim for extension of time. It is thought that, under those circumstances, the costs of carrying out the instruction could be valued under clause 3.6, but if the contractor considers that he has been involved in loss and/or expense, that would be the subject of a common law claim (see Chapter 10).
- The removal or correction of defective work. The clause appears to be broad enough to cover either requirement.
- The postponement of any work in progress. This power does not extend to deferring the giving of possession of the site which would amount to varying an express term of the contract – a matter for the parties to negotiate. Any postponement instruction would inevitably give to an extension of time, a claim for damages at common law and, if the suspension extends to the whole of the works and lasts for a continuous period of at least a month, would provide a ground for determination by the contractor under clause 7.3.1.3 Certain categories of postponement concerning only a part of the works could conceivably be brought under the terms of clause 3.6: 'The Architect ... may order ... a change in the works or the order or period in which they are to be carried out ...', so enabling the architect to include direct loss and/or expense in the valuation.

4.4 Summary

Authority

- From the contractor's point of view, the architect's authority is defined by the terms of the contract.
- The contractor can only take action against the architect in tort

with doubtful chance of success, but the position may be changing.

- The employer can take action against the architect in contract and in tort.
- The architect's authority depends on what he has agreed with the employer.
- The architect is expected to use the same degree of skill as any other average, competent architect unless he professes a higher degree of skill in a particular area.
- Once the contract is in effect, the architect must act fairly to both parties.
- The architect does not have a quasi-arbitral role nor is he immune from actions for negligence as a result of his decisions.
- If the architect is the employee of the employer, he may be in a difficult position.
- The architect cannot overrule the certificates or instructions of a previous architect engaged on the work.
- The architect's approval will generally override contract requirements.
- All instructions must be confirmed in writing by the architect.
- The architect has implied as well as express duties.
- The architect's general power to issue instructions is confined to matters relating to the works.

CHAPTER FIVE
CONTRACTOR

5.1 Contractor's obligations: express and implied

5.1.1 Legal principles

It is all too commonly assumed that the whole of the contractor's contractual obligations are contained and set out in the printed contract form. This is not so and this particular contract form is silent on many important matters. These gaps are filled in by terms which the law will write into the contract.

If nothing at all were said about the contractor's obligations, the law would require him to do three things:

(1) To carry out his work in a good and workmanlike manner exercising reasonable care and skill. This means that the contractor must show the same degree of competence as the average contractor experienced in carrying out that type of work.
(2) To use materials of good quality which are reasonably fit for their purpose.
(3) To ensure that the completed building or structure is reasonably fit for its intended purpose provided that purpose is known. This obligation is modified where the employer engages an architect to design the works since the architect is then responsible for the design.

Similarly, a term would be implied that the contractor would comply with the building regulations and to other statutory requirements – a matter which is the subject of an express term in MW 80.

These implied terms can be modified by the express terms of the contract itself and are, in fact, modified by what MW 80 says, with the result that the contractor is under a lesser duty than would otherwise be the case.

However, statute also imposes similar implied obligations on the contractor and in particular the Supply of Goods and Services Act 1982 implies terms as to the quality of goods supplied by the contractor under the contract. The construction of dwellings (both houses and flats) is governed by the Defective Premises Act 1972 which provides:

> 'Any person taking on work for or in connection with the provision of a dwelling ... owes a duty to see that the work that he takes on is done in a workmanlike manner, with proper materials ... and so as to be fit for the purpose required ...'

The Act envisages a two stage test. Therefore, if the dwelling is fit for habitation despite the fact that some work was not done in a workmanlike manner nor with proper materials, it seems that it complies with the Act: *M.C. Thompson* v. *Clive Alexander & Partners* (1992) CILL 755.

The express provisions of MW 80 must, therefore, be read against this background. There is a further important point. As discussed in Chapter 8, clause 3.2.1 envisages that the contractor may sublet the works or any part of them if the architect gives his written consent. Subcontracting does not free the contractor from responsibility for the subcontracted work. The contractor remains responsible to the employer for all defaults of the subcontractor as regards workmanship, materials or otherwise.

Tables 5.1 and 5.2 summarise the contractor's powers and duties respectively under the express provisions of MW 80.

5.1.2 Execution of the works

Clause 1.1 requires the contractor 'with due diligence and in a good and workmanlike manner (to) carry out and complete the Works in accordance with the Contract Documents ...' which are specified in the first recital. This is his basic and absolute obligation and seems to equate with the common law duty. The contractor is bound to complete the works by the date for completion set out in clause 2.1. He must bring the works to a state where they are practically completed so that the architect can issue his certificate under clause 2.4.

Although the contractor's basic obligation is not qualified in any way in clause 1.1, it is subject to two qualifications:

(1) The provisions of clause 7.3 entitling the contractor to determine his employment under the contract, e.g. if the employer

Table 5.1
Contractor's powers under MW 80

Clause	Power	Comment
2.1	Commence the works on the specified date	
3.1	Assign the contract	If the employer gives written consent
3.2	Sublet the works or part thereof	Only if the architect consents in writing. Consent must not be unreasonably withheld
3.6	Agree the price of variations	<ul><li>If the architect consents *and*</li><li>Before executing the variation instruction</li></ul>
4.8	Suspend performance of obligations under the contract	If employer fails to pay, and has not issued effective notice of withholding or deduction of money, by final date for payment Contractor must first give 7 days notice and must resume work after payment in full
6.4	Reasonably require evidence from the employer that the insurance referred to in clause 6.3B has been taken out and is in force	
7.3.1	Give notice to the employer specifying the default and requiring it to be ended within 7 days	The notice must be served on the employer by registered post, recorded delivery or actual delivery Notice can be served: <ul><li>If the employer fails to pay by the final date for payment the amount properly due *or*</li></ul>

Clause	Power	Comment
Table 5.1 *Contd*		
		• If the employer or any person for whom he is responsible interferes with or obstructs the issue of any certificate or the carrying out of the works *or* • If the employer fails to make the premises available to the contractor on the specified date *or* • If the employer suspends the carrying out of the works for a continuous period of at least one month *or* • If the employer fails to comply with the requirements of the CDM Regulations
7.3.1	Determine the contractor's employment	If default not ended after default notice Determination takes effect on receipt of this notice The notice must not be given unreasonably or vexationally
7.3.2	Determine the contractor's employment	By notice if the employer is insolvent as defined in clause 7.3.2 The notice does not require a prior default notice and it takes effect on receipt by the employer

	Table 5.2 Contractor's duties under MW 80	
Clause	**Duty**	**Comment**
1.1	Carry out and complete the works in accordance with the contract documents in a good and workmanlike manner and with due diligence	If alternative A in the 5th recital applies, in accordance with the Health and Safety Plan of the principal contractor
1.4	Give notice to the Health and Safety Executive required by regulation 7(5) of the CDM Regulations	Where alternative B in the 5th recital applies. The notice must comply with regulations 7(6)(a) and (b) of the CDM Regulations. There must be copies to employer and architect before the contractor begins any construction work
2.1	Complete the works by the specified date	The architect must insert the date
2.2	Notify the architect of delay	If: • It becomes apparent that the works will not be completed by the specified date *and* • This is because of reasons beyond the control of the contractor including compliance with any instruction of the architect whose issue is not due to a default of the contractor
2.3	Pay liquidated damages to the employer	If the works are not completed by the specified date or by any later date fixed under clause 2.2

Table 5.2 *Contd*

Clause	Duty	Comment
2.5	Make good at his own cost any defects, excessive shrinkages or other faults	The defects, shrinkages or other faults must have appeared within 3 months of the date of practical completion *and* must be due to materials or workmanship not in accordance with the contract or to frost occurring before practical completion *and* unless the architect has instructed otherwise
3.3	Keep a competent person in charge upon the works at all reasonable times	Under clause 3.4 the architect may instruct this person's exclusion
3.5	Carry out all architect's instructions forthwith	The instructions must be in writing
4.5.1.1	Supply the architect with all documentation reasonably necessary to enable the final sum to be computed	• Within three months of the date of practical completion • It is not conditional upon a request from the architect
4.5.2	Pay simple interest to the employer on amount not properly paid in the final certificate, until paid	If the contractor fails to pay any amount he is due to pay by the final date for payment
5.1	Comply with and give all notices required by any statute etc.	
5.1	Pay all fees and charges in respect of the works	Provided they are legally recoverable from him
5.1	Give immediate written notice to the architect specifying any divergence	If the contractor finds any divergence between the statutory requirements and the contract documents *or* between such requirements and an architect's instruction

Table 5.2 *Contd*		
Clause	**Duty**	**Comment**
5.7	Comply with all the duties of a principal contractor set out in the CDM Regulations and in this clause	Where the contractor is and remains a principal contractor
5.8	Comply with all the reasonably requirements of the principal contractor	At no cost to the employer where the employer has appointed a successor to the contractor as principal contractor to the extent necessary for compliance with CDM Regulations. No extension of time is to be given
5.9	Provide information reasonably required for the preparation of the health and safety file as required by the CDM Regulations and ensure any subcontractor does the same	Within the time reasonably required in writing by the planning supervisor. If the contractor is not the principal contractor, the information must be provided to the principal contractor
6.1	Indemnify the employer against any expense, liability, loss, claim or proceedings in respect of personal injury or death	If the expense, etc, arises out of or in the course of or is caused by reason of the carrying out of the works except to the extent that it is due to any act or neglect of the employer or those for whom he is legally responsible
6.1	Take out and maintain and cause his subcontractors to maintain the insurances necessary to meet his liability under clause 6.1, including compliance with all relevant legislation	

Table 5.2 *Contd*

Clause	Duty	Comment
6.2	Indemnify the employer against and insure and cause his subcontractors to insure against any expense, liability, loss, claim or proceedings for damage to property other than the works	This is subject to clauses 6.3A or 6.3B. The indemnity operates if the expense etc: • Arises out of or in the course of or is caused by reason of the carrying out of the works *and* • To the extent that it is due to any negligence, omission or default of the contractor or any subcontractor or any person for whom they are legally responsible
6.3A	Insure against the specified risks	New works only
6.3A	Restore or replace work or materials etc., dispose of debris, and proceed with and complete the works	After any inspection required by the insurers following a clause 6.3A claim
6.4	Produce evidence of insurances and cause subcontractors so to do	If the employer reasonably so requires
7.2.3	Immediately cease to occupy the site	Where the employer determines the contractor's employment
7.3.3	Prepare an account setting out: • Total value of work properly executed and materials properly brought on site and other amounts due *and* • Cost to the contractor of moving from site *and* • Direct loss and/or damage	Upon determination of the employment of the contractor under clauses 7.3.1 or 7.3.2. The employer must pay the full amount properly due within 28 days of submission of the account

suspends the carrying out of the works for a continuous period of at least one month, and subject to the procedural requirements of that clause.

(2) If the employer – or the architect on his behalf – prevents the contractor from completing his work by the completion date unless, of course, a proper extension of time has been validly made.

The contractor must carry out his work *in accordance with* the contract documents as defined. The architect must take care to ensure that the description of the work is adequate and this means using precise language. All too often important parts of the contract documents are written in slovenly English. It is impossible to enforce generalisations. The contract documents must contain all the requirements which the employer wishes to impose, and the use of phrases such as 'of good quality' or 'of durable standard' should be avoided. Moreover, the contract documents cannot be used to override or modify the printed conditions because of the last sentence of clause 4.1, the effectiveness of which has been upheld by the courts on many occasions.

The contractor must complete *all* the work shown in, described by or referred to in the contact documents. His obligation only comes to an end when the architect has issued the certificate of practical completion. Thereafter, the contractor must fulfil his obligations under the defects liability clause (see section 9.3.1).

The proviso to clause 1.1 states that if approval of workmanship or materials is a matter for the architect's opinion, then quality and standards must be to his reasonable satisfaction. The effect of this is discussed in Chapter 4, section 4.2.

There is no further amplification of the contractor's basic obligation although he is, of course, bound to complete the works *by* the date stated in clause 2.1 and if he does not do so, subject to the operation of the provisions for extension of time, he must pay liquidated damages to the employer. In *Glenlion Construction Ltd* v. *The Guinness Trust* (1987) 39 BLR 89, it was held, on slightly different wording ('on or before the date for completion'), that the contractor was entitled to complete early, albeit the architect was not obliged to provide information at times to suit early completion. It is suggested that the contractor is similarly entitled under this contract.

The contractor is under a further obligation in progressing the works. He must carry them out with 'due diligence'. The phrase 'regularly and diligently' has been defined in *West Faulkner* v. *London Borough of Newham* (1994) 9 Const LJ 232, by the Court of Appeal:

'Although the contractor must proceed both regularly and diligently with the works, and although each word imports into that obligation certain discrete concepts which would not otherwise inform it, there is a measure of overlap between them and it is thus unhelpful to seek to define two quite separate and distinct obligations.

What particularly is supplied by the word "regularly" is not least a requirement to attend for work on a regular daily basis with sufficient in the way of men, materials and plant to have the physical capacity to progress the works substantially in accordance with the contractual obligations.

What in particular the word "diligently" contributes to the concept is the need to apply that physical capacity industriously and efficiently towards the same end.

Taken together the obligation upon the contractor is essentially to proceed continuously, industriously and efficiently with appropriate physical resources so as to progress the works steadily towards completion substantially in accordance with the contractual requirements as to time, sequence and quality of work.'

We do not believe that the omission of the word 'regularly' is particularly significant in light of this definition. It should be noted that the contractor's failure to proceed with due diligence is one of the grounds which may entitle the employer to bring his employment to an end under clause 7.2.1.

The sensible architect will require a contractor to provide him with a programme, preferably showing logic links, but mere failure to comply with a programme is not usually sufficient alone to show failure to proceed with due diligence. The architect will not approve the contractor's programme, because it is essentially a setting out of the contractor's intentions.

5.1.3 Workmanship and materials

Clause 1.1 also deals with workmanship and materials. It requires the contractor to use materials and workmanship 'of the quality and standards ... specified' in the contract documents. This in fact imposes an obligation on the architect to define the quality and standards of workmanship and materials very carefully indeed. The employer normally places reliance on the architect, not the contractor, to specify correctly and the contractor is unlikely to be liable if he supplies the wrong material merely because the architect has wrongly indicated that any one of a range of materials may be used: *Rotherham Metropolitan Borough Council* v. *Frank Haslam Milan & Co Ltd and Ano* [1996] EGCS 59. Where it is impossible adequately to specify in sufficient detail, the architect could do worse than fall back on the phraseology of the common law:

- All work shall be carried out in a good and workmanlike manner and with proper care and skill.
- All goods and materials shall be of good quality and reasonably fit for their intended purpose.

Quite clearly, the contractor is expected to show a reasonable degree of competence and to employ skilled tradesmen. Although the architect has no power to direct him how he should carry out his work, save to the extent that he can by a variation order issued under clause 3.6 change the order or period in which the works are to be carried out, the architect does have power under clause 3.4 to instruct that any unsatisfactory employee or anyone else employed on the works should be excluded.

Because of the wording of clause 1.1 the contractor *must* provide workmanship, materials and goods of the standards and quality specified. It is no excuse that they are not in fact available, since he does not enjoy the benefit of the limitation of JCT 80 that he need do so only so far as the goods, etc, are procurable. This is a matter which is at the contractor's risk and failure to supply is a breach of contract.

Materials and goods are also referred to in clause 4.2.1 in the context of progress payments. The architect is *bound* to include in progress certificates:

> 'the value of any materials or goods which have been reasonably and properly brought upon the site for the purpose of the works and which are adequately stored and protected against the weather and other casualties'.

This is a very dangerous provision from the employer's point of view since the unfixed materials and goods will not necessarily become the property of the employer, even though he has paid for them, if, in fact, they are not his property in law, as will often the case. Builders' merchants frequently include a 'retention of title' clause in their contracts of sale: *Dawber Williamson Roofing Co Ltd* v. *Humberside County Council* (1979) 14 BLR 70. Ideally, clause 4.2.1

should be amended so as to ensure that the inclusion of the value of unfixed materials is a matter for the architect's discretion. Less satisfactorily, it may include an appropriate provision requiring the contractor's application for progress payments to be accompanied by documentary proof of ownership.

It is probable that the architect is safe from an action for negligence if he does operate clause 4.2.1 as it stands because the employer has signed MW 80 which says that the architect *shall* include on-site unfixed materials. However, architects should be alert to the dangers, and the Joint Contracts Tribunal should issue an amended clause 4.2.1 as a matter of urgency. We first made this plea in the first edition of this book, ten years ago – clearly to no avail.

5.1.4 Statutory obligations

MW 80 clause 5.1 imposes on the contractor a duty to comply with all statutory obligations, e.g. those imposed by the Building Regulations 1991 and subsequent amendments and to pay all fees and charges which are legally recoverable from him. It also imposes on him an obligation to give immediate written notice to the architect if he discovers any divergence between the statutory requirements and the contract documents or one of the architect's instructions.

There then follows a most curious provision:

'Subject to this latter obligation, the contractor shall not be liable to the employer under this contract if the Works do not comply with the statutory requirements where and to the extent that such non-compliance ... results from the contractor having carried out work in accordance with the contract documents or any instruction of the architect.'

The effect of his provision is contractually to exempt the contractor from liability to the employer if the works do not comply with statutory requirements provided he has carried out the work in accordance with the contract documents or any architect's instruction, where, for example, he does not spot the divergence. His obligation to notify the architect of a divergence only arises *if* he spots it, and unless he does so he is under no obligation to notify him. Plainly, it is ineffective to exonerate him from his duty to comply with statutory requirements and if, for example, he carried out work in breach of the Building Regulations, he would be criminally liable since the primary liability to comply with them rests on him: *Perry* v. *Tendring District Council* (1985) 3 Con LR 74. The wording is probably sufficiently wide to protect the contractor from any action by the employer. But it does not protect the architect, and if the fault is his the employer will be able to recover his losses from the negligent architect.

5.1.5 Contractor's representative

Clause 3.3 obliges the contractor to keep on the works a 'competent person in charge' and he must do this at all reasonable times, i.e. during normal working hours. This person is intended to be the contractor's full-time representative on site, but his appointment and replacement are not subject to the architect's approval although in an appropriate case the architect could exercise his powers under clause 3.4 to require his exclusion from the site and thus force the contractor's hand.

Competent means that the contractor's representative must have sufficient skill and knowledge, and it is essential that the architect knows from the outset of this person's identity since he is the contractor's agent for the purpose of accepting instructions. Instructions given to the person in charge are *deemed* to be given to the contractor.

Many architects include in the tender documents a requirement that the contractor give adequate notice of the replacement of the person-in-charge, and this is good contract practice.

5.2 Other obligations

5.2.1 Access to the works and premises

There is no express provision in MW 80 giving the architect and representatives access to the works at all reasonable times. Such a right of access to the site is implied under the general law. However, as is the case under IFC 84, there is a gap because the architect may need access to the contractor's workshops etc. where items are being prepared for the contract. If the architect does need to visit the contractor's premises to keep an eye on things, he must include an appropriate clause in the contract documents, following the wording of JCT 80 clause 11, because it is probable that he does not have that right of access under the general law.

5.2.2 Compliance with architect's instructions

Clause 3.5 obliges the contractor to carry out the architect's instructions forthwith and this obligation is not conditioned in any way. The contractor is given no right to object to architect's instructions, even those involving a variation.

The sanction for non-compliance by the contractor is set out in the second paragraph of clause 3.5. If a contractor fails to comply with an architect's instruction, the architect serves the contractor with a written notice requiring compliance. If the contractor has not complied within 7 days from receipt of that notice, the employer may engage others to carry out the work and deduct the cost from monies due to the contractor. Figure 4.9 is a suggested letter from the architect, and Figure 4.10 is a possible reply.

To avoid the possibility of an argument that the employer has waived his rights, it is essential that the architect ensures that the machinery provided is put into operation if the contractor fails to comply with instructions.

MW 80 contains no express provision for the architect to issue instructions for the removal of work not in accordance with the contract as does JCT 80 clause 8.4, but it is probable that the broad wording of clause 3.5 extends to cover that situation.

5.2.3 Suspension of obligations

Under the provisions of clause 4.8 the contractor has the right to suspend his obligations under the contract under certain circumstances. In effect, this means that the contractor may cease work on site and elsewhere. In order to be able to do so, certain criteria must be satisfied. First, the employer must have failed to pay money due by the final date for payment. Second, the contractor must have given 7 days notice in writing of his intention. The contractor must resume work as soon as he is paid in full. This provision puts the relevant provision in the Housing Grants, Construction and Regeneration Act 1996 into effect. There is no ordinary common law right to suspend work for late payment although, in practice, it is often done. A contractor who suspends work legitimately will be entitled to an extension of time, but not, it seems, loss and expense.

5.2.4 Other rights and obligations

Tables 5.1 and 5.2 summarise the contractor's rights and duties generally. Other matters referred to in those Tables are dealt with in the appropriate chapters.

5.3 *Summary*

Contractor's obligations

MW 80 imposes certain express obligations on the contractor; other obligations are implied at common law:

- The contractor must carry out and complete the works with due diligence and in accordance with the contract documents
- He must do this by the specified completion date
- The contract documents must be precise and specify quality and standards of both workmanship and materials
- The contractor must comply with statutory obligations and pay all fees and charges involved
- *If* he discovers a divergence between the statutory requirements and the contract documents or any architect's instruction he must give the architect written notice immediately
- Subject to this he is exempted from liability to the employer if the works as built contravene statute law
- The contractor must keep a competent representative on site during normal working hours and any instructions given by the architect to that representative are deemed to have been given to the contractor
- The contractor must comply with all architect's instructions issued under the contract
- The employer has an option to engage others to carry out an architect's instructions should the contractor not carry them out within 7 days of a written notice from the architect.
- The contractor may suspend work on 7 days notice if the employer fails to pay.

CHAPTER SIX
EMPLOYER

6.1 Powers and duties: express and implied

Like those of the contractor, some of the employer's powers and duties arise from the express provisions of MW 80, while other arise under the general law.

Tables 6.1 and 6.2 set out those powers and duties of the employer which arise from the express terms of the contract.

This section is concerned with the obligations which are placed on the employer by way of implied terms. These are provisions which the law writes into a contract in order to make it commercially effective. Terms will be implied to the extent that they are not inconsistent with the express terms which may also exclude or modify the implied terms.

It is an implied term of MW 80 that the employer will do all that is reasonably necessary on his part to bring about completion of the contract: *Luxor (Eastbourne) Ltd* v. *Cooper* [1941] 1 All ER 33. Conversely, it is implied that the employer will not so act as to prevent the contractor from completing in the time and in the manner envisaged by the agreement: *Cory* v. *City of London Corporation* [1951] 2 All ER 85. Breach of either of these implied terms which results in loss to the contractor will give rise to a claim for damages at common law and the contractor can pursue that claim under the arbitration agreement.

Equally, if the employer – either personally or through the agency of the architect or that of anyone else for whom he is responsible in law – hinders or prevents the contractor from completion in due time, he is not only in breach of contract, but will be disentitled from enforcing the liquidated damages clause.

The various cases put the duty in different ways, but in essence the position is summarised:

- The employer and his agents must do all things necessary to enable the contractor to carry out and complete the works expeditiously and in accordance with the contract.

Table 6.1
Employer's powers under MW 80

Clause	Power	Comment
2.3	Deduct liquidated damages from any monies due to the contractor under this contract or recover them as a debt	If the works are not completed by the completion date or any extended date and provided that a notice of deduction has been given
3.1	Assign the contract	If the contractor consents
3.5	Employ and pay others to carry out the works Recover all costs incurred from the contractor as a debt	If the contractor fails to comply with a written notice from the architect requiring compliance with an instruction and within 7 days from its receipt
4.4.2	Give written notice to the contractor specifying amount to be withheld or deducted	The notice must state the amount, the grounds and the amount in respect of each ground and must be given not less than 5 days before the final date for payment of a progress payment
4.5.1.3	Give written notice to the contractor specifying amount to be withheld or deducted	The notice must state the amount, the grounds and the amount in respect of each ground and must be given not later than 5 days before the final date for payment of the amount in the final certificate
5.5	Cancel the contract and recover resulting loss from the contractor	If the contractor is guilty of corrupt practices as specified in the clause
6.4	Reasonably require the contractor to produce evidence of insurance referred to in clauses 6.1, 6.2 and 6.3A if applicable	

Table 6.1 *Contd*		
Clause	Power	Comment
7.2.1	Determine the employment of the contractor by written notice	If the contractor • Fails to proceed diligently with the works without reasonable cause *or* • Wholly or substantially suspends the carrying out of the works before completion *or* • Becomes insolvent etc. The determination notice must not be served unreasonably or vexatiously. After the architect's written 7 days default notice except in the case of insolvency

Table 6.2 Employer's duties under MW 80		
Clause	**Duty**	**Comment**
1.3	Immediately notify the contractor in writing of the name and address of the new appointee	If the employee replaces the planning supervisor or the principal contractor pursuant to Article 5.1 or 5.2
4.2.1	Pay to the contractor amounts certified by the architect	Payment must be made within 14 days of the date of the certificate issued under clause 4.2.1
4.2.2	Pay simple interest to contractor on amount not properly paid, until paid	If employer fails to pay any amount due by final date for payment
4.3	Pay similarly	If the architect issues a certificate under clause 4.3
4.4.1	Give written notice to contractor specifying amount to be paid in respect of amount stated as due in any certificate	Not later than 5 days after issue of certificate of payment
4.4.3	Pay the amount stated as due in any certificate	If the employer does not give notices under clauses 4.4.1 and 4.4.2
4.5.1.2	Give written notice to the contractor specifying amount to be paid in respect of amount stated as due in final certificate	Not later than 5 days after issue of final certificate
4.5.1.4	Pay the amount stated as due in the final certificate	If the employer does not give notice under clause 4.5.1.2 and/or clause 4.5.1.3
4.5.2	Pay simple interest to the contractor on amount not properly paid in the final certificate, until paid	If the employer fails to pay any amount he is due to pay by the final date for payment
5.2	Pay to the contractor any VAT properly chargeable	

Table 6.2 *Contd*		
Clause	**Duty**	**Comment**
5.6	Ensure that the planning supervisor carries out his duties Ensure that the principal contractor carries out his duties	If the contractor is not the principal contractor
6.3A	Pay insurance monies to the contractor	On certification by the architect
6.3B	Maintain adequate insurances against the specified risks	Existing structures, contents and the works
6.4	Produce evidence of such insurances	If the contractor so requires
7.3.3	Pay to the contractor the full amount properly due in respect of the contractor's account	If the contractor determines his employment under clauses 7.3.1 or 7.3.2 Payment must be made within 28 days of submission of the account by the contractor

- Neither the employer nor his agents will in any way hinder or prevent the contractor from carrying out and completing the works expeditiously and in accordance with the contract.

The scope of these implied terms is very broad and in recent years more and more claims for breach of them have been before arbitrators and the courts. The employer must not, for example, attempt to give direct orders to the contractor, and he must see that the site is available to the contractor on the date specified in clause 2.1 and that access to it is unimpeded by those for whom he is responsible or over whom he has effective control. For example, in *Rapid Building Co Ltd v. Ealing Family Housing Association Ltd* (1985) 1 Con LR 1, squatters had occupied part of the site and as a result the employers were unable to give possession on the due date. This was a breach of contract which caused appreciable delay to the contractors who were held entitled to damages. This is especially important in the case of work to existing structures or occupied buildings, and these are matters which should be discussed between the employer and his professional advisers before the contract is let.

6.2 Rights under MW 80

6.2.1 General

Although the contract is between the employer and the contractor – who are the only parties to it – an analysis of the contract clauses shows that the employer has few express rights of any substance.

His major right is to have the work contracted for handing over to him by the agreed completion date, properly completed in accordance with the contract documents. He also has the right to assign the benefits of the contract to a third party, provided the contractor consents in writing (clause 3.1), although it is difficult to envisage many circumstances in which an employer could wish to do that!

6.2.2 Damages for non-completion

If the contractor does not complete the works by the agreed completion date, the employer is entitled to recover liquidated damages at the rate specified in clause 2.3 for each week or part of a week during which the works remain uncompleted after the original or extended completion date.

There is no requirement that the architect should issue a certificate of non-completion. The mere fact of late completion is sufficient to bring clause 2.3 into operation. The employer is now given an express right to deduct liquidated damages from monies due to the contractor, but if (unusually) no sums are due or to become due to the contractor, the employer must sue for them as a debt.

Amendment MW11 requires that before any money can be deducted from money due to the contractor, the employer must have given written notice under clauses 4.4.2 or 4.5.1.3 to comply with the Housing Grants, Construction and Regeneration Act 1996. Inexplicably, the employer must also inform the contractor in writing not later than the date of the issue of the final certificate. This duty is stated to be additional. Therefore, if the employer, unusually, waits until the final certificate to deduct liquidated damages, he must serve one notice not later than 5 days before payment is finally to be made and an earlier notice when the final certificate is issued. Were it not for the use of the word 'additionally', a single notice on the issue of the final certificate would have been enough to satisfy the provision.

In practice, the architect should, of course, advise the employer of his rights should the contract overrun.

6.2.3 Other rights

These are summarised in Table 6.1 and are there described as 'powers'. They are discussed in the appropriate chapters.

6.3 *Duties under MW 80*

6.3.1 General

The essence of a duty is that it is something that must be done. It is not permissive; it is mandatory. Breach of a duty imposed by the contract will render the employer liable to an action for damages by the contractor in respect of any proven loss.

Some breaches of contract may, in fact, entitle the contractor to treat the contract as at an end. Lawyers call them 'repudiatory breaches' which means that they go to the basis of the contract. For example, physically expelling the contractor and his men from site would be a repudiatory breach since the employer is effectively showing that he no longer wishes to be bound by the contract.

Alternatively, for the contractor to walk off site permanently before the works are complete would be a repudiatory breach on his part.

Breach of any contractual duty will always, in theory, entitle the contractor to at least nominal damages, although in some cases any loss will be difficult if not impossible to quantify.

6.3.2 Payment

From the contractor's point of view, the most fundamental duty of the employer is to make payment in accordance with the terms of the contract. However, while steady payment against certificates is essential from the contractor's viewpoint, the general law does not regard non- or late payment as such as a major breach of contract. Certainly non-payment of one certified sum to the contractor is not repudiation, but in *D.R. Bradley (Cable Jointing) Ltd* v. *Jefco Mechanical Services Ltd* (1988) 6-CLD-07-21, a subsequent refusal to pay on a certificate did go to the root of the contract since 'it reasonably shattered [the contractor's] confidence in being paid'. Case law also suggests that a certificate is not as good as cash, as most contractors appear to think, since in some limited cases the employer may be justified in withholding payment of certificated amounts pending arbitration: *C.M. Pillings & Co Ltd* v. *Kent Investments Ltd* (1986) 4 Con LR 1; *R.M. Douglas Construction Ltd* v. *Bass Leisure Ltd* (1991) 25 Con LR 38. That is particularly so since Amendment MW11 has introduced specific notices to be given if the employer wishes to withhold payment or deduct any amounts in respect of an interim progress payment (clause 4.4.2) or the final payment (clause 4.5.1.3). The employer must have a substantial reason for withholding payment and he must indicate the grounds and the amount proposed to be withheld in respect of each ground, otherwise the contractor may be justified in suspending work under clause 4.8 or giving a default notice prior to determination under clause 7.3.1. Clauses 4.2.2 and 4.5.2 now contain provisions requiring the employer to pay simple interest on amounts not properly paid.

MW 80 is quite specific in its terms as to payment (see Chapter 11). The basic provisions are to be found in clauses 4.2, 4.3, 4.4 and 4.5.

If the employer is a tardy payer, he should be made aware of the need to pay promptly in accordance with the contract terms because nothing sours good working relationships more than late payment, and most contractors do have a cash-flow problem. In these circumstance the architect ought to write to the employer pointing out the contractual position (Figure 6.1).

**Figure 6.1
Letter from architect to employer if employer is
slow to honour certificates**

Dear Sir

PROJECT TITLE

The contractor has complained to me that he is not receiving payment
due on certificates until after the period allowed in the contract.

Under the terms of the contract, you have fourteen days from the date
of the certificate within which to make payment. Failure to pay within
the stipulated time entitles the contractor to suspend his obligations
under the contract or determine his employment under the contract. If
the contractor determined his employment, the consequences would
be extremely expensive. Sums outstanding attract simple interest at
the rate of 5% above Bank of England Base Rate.

Disregarding the strict legal requirements as to payment, it is good
practice to pay promptly because the contractor is always in the
position of having paid out substantial sums well before payment is
due. Prompt payment is crucial to his cash flow and, consequently, late
payment spoils good working relations.

If you would bear this in mind, there should be benefits on all sides.

Yours faithfully

Once the architect has issued a payment certificate under clauses 4.2 and 4.3 (progress payments and penultimate certificate) the employer is given a period of grace in which to honour the certificate. He must make payment *within 14 days of the date of the certificate*, which means payment before the expiry of that period. This a very tight timetable and the architect should send the certificate to the employer on the day it is issued. Assuming no postal delays (which cannot be relied on these days) and dispatch by first-class post, the employer effectively has no more than 13 days for paying what is due. The employer must give notice in writing (clause 4.4.1) to the contractor within five days from the date of issue of a certificate. The notice must state the amount of payment to be made in respect of the amount stated as due in the certificate. This appears to give the employer the chance to abate the certified sum, because the second notice the employer may give relates to any amount withheld from the amount stated in the first notice (see clause 4.4.2). However, it should be noted that the contractor may suspend work if the employer does not pay the amount certified subject only to the second notice (clause 4.4.2). The determination provision in clause 7.3.1.1 is less precise, simply referring to a failure to 'pay by the final date for payment the amount properly due to the Contractor in respect of any certificate . . .', leaving open the question of whether the amount properly due is the same as the certified sum.

The same timetable effectively applies to the final certificate under clause 4.5, though the wording in that case is that the 'final date for payment by the Employer of the amount so certified shall be 14 days from the date of issue of the certificate.'

Payment by cheque is probably good payment, but it is no excuse for the employer to say, for example, that his computer arrangements do not fit in with the scheme of certificates, which is an increasingly common excuse for late payment. If this is indeed the case, the payment period should have been amended before the contract was let.

The employer is entitled to retain retention on progress payments and, in the case of a contract overrun, can set off amounts due to him as liquidated damages under clause 2.3 provided the complex notice provisions are satisfied.

6.3.3 Retention

The employer has certain rights in the retention percentage, which is commonly 5%. It is not stated that the retention is trust money and it may be, therefore, that the contractor is at risk if the employer becomes insolvent.

The employer's rights in the retention are to have it as a fund from which he is entitled to have defects remedied and other bona fide claims settled. But morally the retention is the contractor's money and we once again suggest that it would be an improvement if the JCT would amend clause 4.2 to provide that the employer's interest in the retention was 'fiduciary as trustee for the contractor'.

6.3.4 Other duties

These are summarised in Table 6.2 and are commented on as appropriate in other chapters.

6.4 Summary

Rights and duties

- The employer must not hinder or prevent the contractor from carrying out and completing the works as envisaged by the contract. Breach of this duty will render the employer liable to an action for damages and may disentitle him from enforcing the liquidated damages clause.
- Should the contractor fail to complete the works by the completion date, the employer is entitled to recover liquidated damages.
- The employer must pay the contractor on the architect's certificates, payment being due within 14 days of the date of each certificate.
- The retention fund may be used by the employer to satisfy bona fide and quantified claims but it is the contractor's money and not the employer's.

CHAPTER SEVEN
QUANTITY SURVEYOR AND CLERK OF WORKS

7.1   *Quantity surveyor*

7.1.1   Appointment

MW 80 makes provision for the appointment of a quantity surveyor in Recital 4 which contains the only reference to the quantity surveyor in the entire contract. There is a footnote which directs the architect to 'Delete as appropriate', i.e. if no quantity surveyor is appointed.

The decision whether or not to appoint a quantity surveyor will be taken by the employer with the architect's advice. His advice will naturally take into account the amount of work which the quantity surveyor could be asked to carry out. In the case of small works, the use of a quantity surveyor tends to be the exception. The architect should be competent to deal with interim payments, the computation of the final sum and valuation of variations if the work is of a simple and straightforward nature, particularly if a clause is inserted to provide for stage payments. If, however, the architect has any doubts about his competence in this field or the work is complex, the use of a quantity surveyor is indicated. It is important to remember that, if the architect holds himself out to his client as capable of carrying out quantity surveying functions, that is the standard which will be expected of him. The architect should not fill in his own name as quantity surveyor unless he has checked his position with his professional indemnity insurers. Since there is no further mention of the quantity surveyor in the contract, there appears to be no valid reason why the architect should ever insert his own name. There is no benefit in so doing, and there may be insurance problems.

If the architect decides that a quantity surveyor is required for the works, he should inform the employer at the earliest possible opportunity, usually stage A – Inception, so that the quantity surveyor can assist throughout the project. The architect should take care to explain to the employer why a quantity surveyor is required (Figure 7.1). The situation will be simplified if the employer has been given a copy of the RIBA Conditions of Engagement CE/95 on which, presumably, the architect's engagement is based. It is worth noting at this point that the architect may be engaged under the Standard Form of Agreement for the Engagement of an Architect (SFA 92), which is an earlier and somewhat more complex set of terms.

7.1.2 Duties

The contract goes no further than to provide for the appointment of a quantity surveyor; it is silent regarding his duties. Although it is probable that an implied term would govern the matter, if a quantity surveyor is appointed it is best to deal with his powers and duties by means of a suitably worded provision which gives him a right of access to the site. There are two possible ways of dealing with this:

(1) The insertion of a brief clause in the contract to cover the activities the architect intends the quantity surveyor to perform
(2) A letter to the contractor informing him that the quantity surveyor is to be the architect's authorised representative in respect of specified activities.

The introduction of a clause in the contract is probably the more satisfactory way of accomplishing the objective. The contractor then knows, at tender stage, just what is intended. If the architect handles the matter by means of a letter at the beginning of the contract, the contractor might object violently to it and relations will have been soured at the start. Another danger is that the contractor may not appreciate the limits of the quantity surveyor's duties and may carry out as instructions what are simply the quantity surveyor's comments on some aspect of the work. Although the contractor would be wrong to do so, it is little consolation to the employer if disruption and delay results. If it is felt that the quantity surveyor's duties must be dealt with in this way, great care must be taken with the letter (Figure 7.2).

If a clause is included, proper legal assistance must be obtained for its drafting. Among the duties that the architect wishes the quantity surveyor to carry out may be the following:

Figure 7.1
Letter from architect to employer if the services of a quantity surveyor are required

Dear Sir

PROJECT TITLE

In view of the particular nature/size/value [*delete as appropriate*] of this project, the services of a quantity surveyor will be required. He is the specialist in building economics and he will deal with preparation of valuations for progress payments and variations and provide overall cost advice. He should be appointed at this stage so that you can derive the maximum benefit from his advice and he can become fully involved with the project.

You may wish to consider the use of [*insert name*] of [*insert address*], with whom I have worked many times in the past. If you will let me have your agreement, I will carry out some preliminary negotiations on your behalf and advise you regarding the letter of appointment.

The use of consultants is covered by clauses 4.1.1 to 4.1.9 inclusive of the Conditions of Engagement CE/95, a copy of which is already in your possession.

Yours faithfully

Figure 7.2
Letter from architect to contractor regarding the duties of the quantity surveyor

Dear Sir

PROJECT TITLE

The quantity surveyor named in Recital 4 of the contract is [*insert name*] of [*insert address*]. This letter is to notify you that he will be my authorised representative to carry out the duties of valuation, calculation, checking, computation and measurement in respect of the following clauses of the Conditions of Contract:

2.5
3.5
3.6
4.2
4.3
4.4.2
4.5
4.6
6.3A/6.3B [*delete as appropriate*]
7.2.3.1
7.3.3

You must supply him with all necessary documents, vouchers, etc. to enable him to carry out his work. Your attention is drawn to the fact that the quantity surveyor's duties are limited to quantification. He is not empowered to issue instructions, certificates, make awards or decide liability for payment.

Yours faithfully

Copies: Employer
          Quantity surveyor

- Value the amount of the employer's contribution under clause 2.5, if the architect instructs otherwise than that the contractor should make good all defects at his own expense (see section 9.3.4)
- Valuation work under clause 3.5 if the employer has to employ others to carry out work contained in architect's instructions
- Valuation of variations in accordance with clause 3.6
- Measurement and valuations under clause 4.2
- Measurement and valuations under clause 4.3
- Calculations required for clause 4.4.2
- Computation of the final sum under clause 4.5
- Calculations under clause 4.6
- Calculations under clauses 6.3A and 6.3B
- Valuations under clause 7.2.3
- Checking the contractor's account prepared under clause 7.3.3.

Provision should be included for the quantity surveyor to require the contractor to supply any necessary documents for the purpose of carrying out the valuations, calculations or measurements. It is also prudent to stress that the quantity surveyor's duties cover quantification only with no power to agree or decide liability for payment on or to issue instructions, certification or make any awards and this would be the situation under the general law: *John Laing Developments Ltd* v. *County and District Properties Ltd* (1982) 23 BLR 1.

7.1.3 Responsibilities

The quantity surveyor's responsibility is to the employer in contract and in tort: *Wessex Regional Health Authority* v. *HLM Design and Webb* (1994) 10 Const LJ 165. Even if a list of his duties is included in a special clause inserted in MW 80, he will not be liable to the contractor in contract because, like the architect, he is not a party to the contract. There is a remote possibility that he might be liable to the contractor in tort but the architect is the professional entrusted with the task of certifying all payments. If the quantity surveyor makes a mistake, the architect must correct it. The onus on him to satisfy himself about all the quantity surveyor's calculations cannot be emphasised too much.

Clause 4.1.7 of the RIBA Conditions of Engagement CE/95 states that, where the client employs a consultant, he will hold that consultant responsible for the competence, general inspection and

103

performance of the work entrusted to that consultant. The fore-runner of this clause in the RIBA Conditions of Engagement was upheld in *Investors in Industry Commercial Properties Ltd* v. *South Bedfordshire District Council* (1986) 5 Con LR 1 where the Court of Appeal said that following the appointment of a consultant by the client on the architect's recommendation:

> 'the architect will normally carry no legal responsiblity for the work done by the expert which is beyond the capability of an architect of ordinary competence ... [but] this is subject to one important qualification. If any danger or problem arises in connection with the work allotted to the expert of which an architect of ordinary competence reasonably ought to have been aware and reasonably could be expected to warn the client despite the employment of the expert, and despite what the expert says or does about it, it is ... the duty of the architect to warn the client about it.'

However, if the quantity surveyor makes a mistake, the architect will always be the first target for the client's displeasure. Moreover, clause 4.1.9 goes on to state that nothing in the clause will affect the architect's responsibility for issuing instructions or for other functions ascribed to him under the contract. One of the 'other functions', of course, is the certification of payments to the contractor.

If the quantity surveyor gives advice directly to the employer regarding any matter, he is liable if the advice is negligently given. In practice, this will happen only rarely. An example might be if the quantity surveyor is present at the meeting to open tenders for the main contract and volunteers negligent advice, as a direct result of which the employer enters into a contract with a contractor which turns out to be more expensive than the employer was led to believe. Even in this sort of instance, there is a very good chance that the employer will blame the architect for the difficulty. The employer may have a point, but everything will turn on the precise circumstances in which the quantity surveyor gave his advice. The quantity surveyor has exactly the same sort of contractual relationship with the employer as the architect has. The architect's duties are different and he has, in addition, the authority and duty to co-ordinate and integrate the consultant's services (Conditions of Engagement CE/95, clause 4.1.5), but the architect can never be responsible for the quantity surveyor's actions or defaults.

The architect should resist, as far as possible, any efforts on the part of the employer to get him to appoint the quantity surveyor himself. From the employer's point of view it is understandable that he wishes to deal with everything through one person, but the arrangement leaves the architect very exposed. The modern practice in negligence cases seems to be to sue everyone in sight and some commentators argue that it makes no real difference whether the employer appoints all consultants directly or simply appoints the architect and leaves him to appoint any other consultants, with the employer's permission, who may be necessary. We do not share that view. Figure 7.3 indicates the various relationships which are possible. If the employer suggests that the architect appoint the quantity surveyor, we suggest that he put his position on record (Figure 7.4).

It should be remembered that the architect has no contractual relationship with the quantity surveyor (if the employer has appointed him); if there is any liability between architect and quantity surveyor it will be in tort.

7.2 Clerk of works

7.2.1 Appointment

MW 80 makes no provision for a clerk of works. This is because, quite clearly, on most works for which MW 80 is suitable, the employment of a clerk of works in a full or part-time capacity is hardly justified. Every project has its own difficulties, however, and if it is considered that a clerk of works is required, the architect must advise the employer accordingly.

The Conditions of Engagement CE/95 expressly state in clause 3.1.1 that the architect will make such visits as he reasonably thought would be necessary at the date of appointment. If it becomes clear that more frequent or even constant inspections are required, the architect may either recommend that he carries them out himself at an adjusted fee rate (clauses 3.1.2 and 3.1.3) or recommend the appointment of a clerk of works (clause 3.3). If the employer is one of those organisations which employ clerks of works on their permanent staffs, that is an excellent arrangement. To avoid misunderstandings, the position should be put on record (Figure 7.5). In the case of most employers, however, it will be a one-off arrangement, the clerk of works being engaged by the employer on a full-time or part-time basis for the duration of the contract. Some firms of architects employ their own clerks of works on a permanent basis, but, in the light of case law, the architect is advised

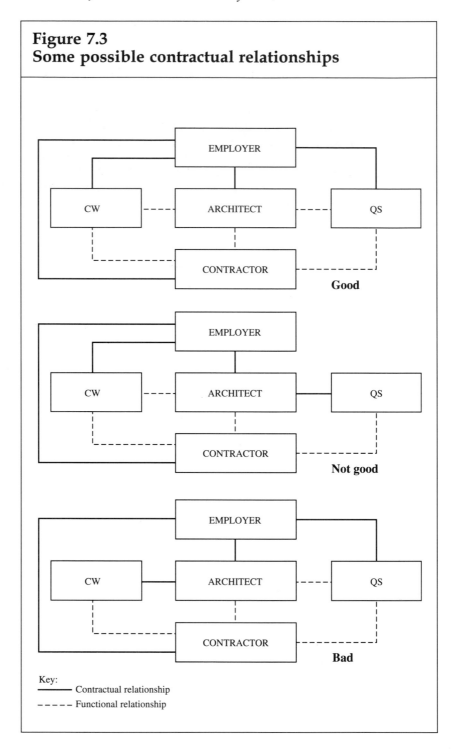

Figure 7.3
Some possible contractual relationships

Figure 7.4
Letter from architect to employer if he requires the architect to appoint the quantity surveyor

---

Dear Sir

PROJECT TITLE

Thank you for your letter of the [*insert date*] and I am pleased that you have agreed to the appointment of [*insert name*] as quantity surveyor for this project.

I strongly advise you to appoint the quantity surveyor directly yourself. That is normal practice in construction projects and gives you direct right of access to him if you should so wish. I cannot provide these services myself and, if you wish me to appoint, I should do so simply as your agent.

I will, in any case, co-ordinate all professional work. The quantity surveyor's fees are the same whether appointed directly, as I advise, or through my office. If you let me know that you will take my advice, I will draft an appropriate letter of appointment for you to send.

Yours faithfully

Figure 7.5
Letter from architect to employer regarding the appointment of a clerk of works

Dear Sir

PROJECT TITLE

In view of the nature/size/value [*delete as appropriate*] of this project, I advise the appointment of a clerk of works on a part-time basis, say [*insert number*] hours per week.

The RIBA Conditions of Engagement CE/95, clause 3.12, provides that if it becomes apparent that my expectation of site visits needs to be varied, I should inform you of my recommendations. A clerk of works will be able to make inspections of such frequency as should ensure the proper carrying out of the work. He is likely to repay his cost several times over in savings on lost time and money as the contract progresses.

Although naturally, I will carry out my own duties with proper skill and care, on a contract of this type I cannot accept responsibility for such defects as would be discovered by the employment of a clerk of works.

Yours faithfully

to avoid employing a clerk of works himself, even though the employer pays him an additional sum to cover the clerk of works' salary (see section 7.2.3).

If a clerk of works is to be used, it is essential to include a clause to that effect in the contract. It is suggested that the clause follows the lines of clause 3.10 of IFC 84 Intermediate Form which simply states:

'The Employer shall be entitled to appoint a clerk of works whose duty shall be to act solely as an inspector on behalf of the Employer under the directions of the Architect.'

This states all that is necessary and removes any uncertainty in the contractor's mind regarding the directions of the clerk of works – he is not empowered to issue any. If he does, the contractor should write to the architect accordingly (Figure 7.6).

In the interests of the project, the architect should advise the employer to appoint the clerk of works as soon as the contractor's tender has been accepted. This gives the clerk of works time to thoroughly familiarise himself with the drawings, specification and schedules. The architect will hold a meeting with him to brief him about the work to be done, but the architect would be wise to confirm the main points to the clerk of works by letter (Figure 7.7) which should outline the aspects of his job considered to be important and remind him of the extent and limitations of his duties.

7.2.2 Duties

If a clause is incorporated based upon IFC 84, clause 3.10, the clerk of works' sole duty is to inspect the works. The clause makes clear that he carries out his function on behalf of the employer; he is not the architect's inspector. This is vitally important as will be seen in the next section. The employer, however, has no power to direct him. That is the architect's prerogative alone.

The architect's directions to the clerk of works will presumably embrace how, where and at what intervals he should inspect and to what he should pay particular attention. If the architect issues any directions which are other than purely routine in nature, he should always confirm them in writing to protect his position. In practice, the clerk of works will often do more than simply inspect. He will usually be a person of some experience whose advice the architect will seek from time to time. The contractor often asks the clerk of

---

Figure 7.6
Letter from contractor to architect regarding directions from the clerk of works

---

Dear Sir

PROJECT TITLE

The clerk of works has issued direction number [*insert number*] dated [*insert date*] on site, a copy of which is enclosed.

Such directions have, of course, no contractual effect. Clearly, the directions of the clerk of works issued in relation to the correction of defective work can be very helpful. We are anxious to avoid misunderstandings on site and in this spirit we suggest that the clerk of works should issue no further directions, other than those relating to defective work. There is no reason why he cannot refer all other matters directly to you by telephone and, if appropriate, you can issue an architect's instruction as empowered by the contract.

In our view, the above system would remove a good deal of the uncertainty which must result from the present state of affairs. We look forward to hearing your comments.

Yours faithfully

Figure 7.7
Letter from architect to clerk of works setting out duties

Dear Sir

PROJECT TITLE

My client [*insert name*] has confirmed your appointment as clerk of works for the above contract. I should be pleased if you would call at this office on [*insert date*] at [*insert time*] to be briefed on the project and to collect your copies of drawings, specification, schedules, weekly report forms and site diary.

The contractor is expected to take possession of the site on the [*insert date*]. You will be expected to be present on site for a minimum of [*insert number*] hours per week from that date. I will discuss the timing of your visits when we meet. Let me know at the end of the first week if proper accommodation is not provided for you as described in the specification.

Your duties will be as indicated in the Conditions of Contract special clause [*insert number*], a copy of which is enclosed for your reference. In particular, I wish to draw your attention to the following:

(1) You will be expected to inspect all workmanship and materials to ensure conformity with the contract, i.e. the drawings, specification, schedules and further instructions issued from this office. Any defects must be pointed out to the person in charge, to whom you should address all comments. If any defects are left unremedied for twenty-four hours or if they are of a major or fundamental nature, you must let me know immediately by telephone.

(2) Although it is common practice for clerks of works to mark defective work on site, you must not make such marks or in any way deface materials on site.

(3) It is not my policy to issue lists of defects to the contractor before practical completion (commonly known as 'snagging lists'). They are open to misinterpretation and should be compiled by the person in charge. Confine yourself to oral comments.

(4) The architect is the only person empowered to issue instructions to the contractor.

Figure 7.7 *Contd*

---

(5) Any queries, unless of a minor explanatory nature, should be referred to me for a decision. You are not empowered to vary or omit work.

(6) The report sheets must be filled in completely and a copy sent to me on Monday of each week. Pay particular attention to listing all visitors to site and commenting on work done in as much detail as possible.

(7) The diary is provided for you to enter your daily comments.

(8) Remember that your weekly reports and site diary may be called in evidence should a dispute arise, so you must bear this in mind when making your entries.

The successful completion of the contract depends in large measure on your relationship with the contractor. If you are in any doubt about anything, please let me know.

Yours faithfully

works for assistance in solving site problems. Note, however, that, if the contractor acts on the advice or even instructions of the clerk of works, he does so at his peril. In giving advice, the clerk of works is acting in a personal capacity.

The architect should make sure that the contractor understands the position at the beginning of the contract. It is prudent to give some time to this topic at the first contract meeting and give some space to it in the minutes. Contractors get into the habit of accepting the clerk of works as speaking on behalf of the architect. Although there is no foundation for this view in the contract, it saves much bad feeling to put the matter beyond doubt. One of the worst things that can happen on a contract is for the architect to have to overrule the clerk of works.

The duties of the clerk of works can, thus, be seen in two parts:

(1)   His duties under the contract
(2)   His duties by virtue of the architect's specific directions to him.

His duties under the contract define his relations with the contractor and his duty to comply with the architect's directions defines his relations with the architect. That is to put the matter in broad terms. Every architect has his own views about his relationship with the clerk of works, but among the duties he might be expected to fulfil are the following:

- Inspect the works
- Relay queries and problems back to the architect
- Complete report sheets
- Complete daily diary
- Take measurements as directed
- Take particular notes of such things as portions of the work opened up for inspection under clause 3.5.

The clerk of works is not empowered to put any marks on defective portions of the work. Once such work is removed, it is the property of the contractor who is entitled to expect that the clerk of works will not do any damage to the works, no matter how slight. If this problem does occur, the contractor should write to the architect immediately (Figure 7.8).

It is, unfortunately, common practice for the clerk of works to issue 'snagging lists' to the contractor, particularly towards the end of a job. There are very rare occasions when there may be pressing

Figure 7.8
Letter from contractor to architect if the clerk of works defaces work or materials

Dear Sir

PROJECT TITLE

It is common practice for the clerk of works to deface work or materials which he considers to be defective. The basis for such action is to bring the defect to the notice of the contractor and ensure that it cannot remain without attention.

We object to the practice because:

(1) The work or materials so marked may not be defective and we will be involved in extra work and the employer in extra cost in such circumstances.
(2) The work or materials so marked, if indeed defective, will not be paid for and will be our property when removed. We may be able to incorporate it in other projects where a different standard is required. Defacement by the clerk of works would prevent such reuse.

We will take no point about the defacing marks we noted on site today, but if the practice continues, we will seek financial reimbursement on every occasion.

Yours faithfully

reasons for taking this course of action. Generally, the practice should be discouraged because:

- The clerk of works is inspecting for the benefit of the employer, he owes no duty to the contractor to find defects
- It is the job of the person in charge to produce his own lists of work
- The contractor may be under the impression, however misguided, that if he simply attends to defects on the 'snagging lists', his obligations are at an end, and disputes may follow.

Obviously, the clerk of works must draw the contractor's attention to work not in accordance with the contract documents, but he should not be more specific. In particular, neither he nor the architect should instruct how defective work is to be corrected if the only problem is that it is not in accordance with the contract. To do otherwise may result in the employer having to pay for the work: *Simplex Concrete Piles Ltd* v. *Borough of St Pancras* (1958) 14 BLR 80.

7.2.3 Responsibilities

Like everyone else connected with the contract, the clerk of works has a responsibility to carry out his duties in a competent manner. He must demonstrate the same degree of skill that would be demonstrated by the average clerk of works. If he holds himself out to be especially skilled in some branch of his work, a greater standard of skill than the average will be expected in that particular branch.

Case law has made clear that, provided the ordinary relationship of master and servant exists between employer and clerk of works, the employer will be vicariously liable for the actions of the clerk of works in the normal way: *Kensington and Chelsea and Westminster Area Health Authority* v. *Wettern Composites and Ors* (1984) 1 Con LR 114; *Gray* v. *T.P. Bennett & Son* (1987) 43 BLR 63. The fact that the clerk of works is under the architect's directions makes no difference. The relationship between the clerk of works and architect has been compared to that between the chief petty officer and the captain of a ship. This does not mean that, if the clerk of works is negligent, it will relieve the architect of all responsibility, but it may substantially reduce his liability for damages depending on circumstances. It is, therefore, very much in his interests to ensure that the employer employs a clerk of works. If the architect employs the clerk of works, his negligence will not reduce the architect's liability.

7.3 *Summary*

The quantity surveyor

- The contract provides for the appointment of a quantity surveyor, but not his duties.
- It is for the architect to advise the employer on the appointment.
- There is a danger in the architect holding himself out as capable of carrying out quantity surveying functions.
- If the architect acts as quantity surveyor, he must check with his professional indemnity insurers.
- A clause should be inserted in the contract or a letter written to the contractor to cover his duties.
- His duties should include deciding quantum, but not liability to pay.
- The quantity surveyor should be appointed by the employer, to whom he will be liable.
- It is the architect's responsibility to check that all certificates are correct.
- Despite legal liabilities, the employer will probably look to the architect first if something goes wrong.

The clerk of works

- No provision in the contract.
- The architect should advise the employer if he thinks a clerk of works is necessary.
- A suitable clause should be inserted in the contract.
- The clerk of works should be appointed by the employer.
- He should be appointed immediately the successful tender has been accepted.
- He should be briefed thoroughly.
- Sole duty should be to inspect the works.
- He may carry out other duties for the architect.
- He must not put marks on the work.
- He should not issue 'snagging lists'.
- He should not instruct how defective work is to be made good in accordance with the contract.
- He must be competent.
- His employment may reduce the architect's liability for damages if the employer is found to be vicariously liable.

CHAPTER EIGHT
SUBCONTRACTORS AND SUPPLIERS

8.1 General

This chapter deals with third parties insofar as they are provided for or might affect the work carried out under MW 80. Subcontractors, suppliers, statutory authorities, persons engaged directly by the employer and the possibility of nominating or naming sub-contractors are considered.

The contract provisions are extremely brief. They are contained in clauses 3.1, 3.2 and 5.1. There are no provisions for suppliers, employer's licensees or nominated or named subcontractors.

8.2 Subcontractors

8.2.1 Assignment

Assignment is usually coupled with subcontracting in contract provisions and MW 80 is no exception. It is a mistake, however, to consider that they are linked. They are totally different concepts. Assignment is the legal transfer of a right or duty from one party to another whereby the original party retains no interest in the right or duty thereafter. For this to be fully effective *novation* must take place, that is, the formation of a new contract.

Subcontracting is, in essence, the delegation of a duty from one party to another, but the original party still retains primary responsibility for the discharge of that duty. It is vicarious performance of a duty by someone else.

Clause 3.1 deals with assignment. Both employer and contractor are prohibited from assigning the contract unless one has the written consent of the other. This is a much stricter provision than is to be found under the general law where it is possible for either party to assign the benefits or rights of a contract to a third party. For example, it is quite common for the contractor to wish to assign to a third party the benefit of receiving progress payments in return for substantial financial help at the beginning of the contract. The employer, too, may wish to sell his interest in the completed building to another before the issue of the final certificate, thus assigning the benefit of the contract. Although permitted under the general law, clause 3.1 operates to stop such deals. Duties or burdens of contracts can never be assigned without express agreement between the parties. The effectiveness of this kind of clause was considered by the House of Lords in *Linden Gardens* v. *Lenesta Sludge Disposals* and *St Martin's Property Corporation* v. *Sir Robert McAlpine & Sons* (1993) 9 Const LJ 322. The clause was held to be effective to prevent assignment of the benefits of a contract or the right to damages for breach of the contract.

It is considered that the architect has a duty to explain the possible difficulties to the employer, particularly if he has reason to believe that the employer might wish to assign his benefit before the final certificate is issued. Assignment can be made with the written consent of the other party, but such consent may be refused and the grounds for refusal need not be reasonable. It would appear to do no harm, and make much sense, for the architect to advise the employer to amend the contract so as to prohibit only the assignment of duties under the contract. The amendment should be carried out at tender stage and should result in lower rather than higher tenders.

8.2.2 Subcontracting

Clause 3.2 deals with subcontracting. Subcontracting is traditional in the building industry, but the practice can be abused to the extent that the contractor's sole employee on the site might be the person-in-charge, the remainder of the workforce being subcontractors. Needless to say, such an arrangement does not make for an efficient contract. Clause 3.2.1 is designed to prevent this and other problem situations by allowing subcontracting only if the architect gives his written consent (see Figure 8.1). Although he may withhold his consent, he must act reasonably. It is probably reasonable for him to require the contractor to supply him with the names of the proposed subcontractor before he consents.

It is important to remember that there is no contractual relationship between the employer and the subcontractor. The subcontractor's contract is with the contractor. It follows that nothing contained in MW 80 is binding in any way on the subcontractor. It is vital, therefore, that the subcontract gives the con-

118

Figure 8.1
Letter from contractor to architect requesting consent to subletting

---

Dear Sir

PROJECT TITLE

We propose to sublet portions of the works as indicated below because [*state reasons*]. We should be pleased to receive your consent in accordance with clause 3.2.1.

[*list the portions of the works and the names of the subcontractor*]

Yours faithfully

tractor sufficient controls over the subcontractor because the employer must look to the contractor for redress if the subcontractor defaults.

There is no standard form of subcontract for use with MW 80. It is thought that the architect would not be unreasonable if he withheld his consent to subletting until he had satisfied himself that the contractor's subcontract provisions were adequate. He should beware, however, of being drawn into disputes between contractor and subcontractor or of being seen to approve the form of subcontract. It should be remembered that ultimately the contractor is responsible for his own subcontractors and he has no right to look to the architect or the employer to assist him if things go wrong.

The lack of detailed provisions is sometimes to be regretted, but is unavoidable taking the contract as a whole. Unfortunately, although the number of subcontractors will tend to increase with the size of the work, there is no lower point at which the contractor will not use any subcontractors at all. Clause 3.2.2 stipulates that a subcontract must include a provision which makes the contractor liable to pay simple interest to a subcontractor if the contractor fails to pay the amount due to the subcontractor by the final date for payment. The rate of interest is to be 5% over the Bank of England Base Rate as in clauses 4.2.2 and 4.5.2 where late payment to the contractor is concerned. The interest is to be treated as a debt owing from the contractor to the subcontractor.

There is a further provision that payment of the interest is not to be construed as a waiver by the subcontractor of his right to proper payment. In other words, just because there is a rate stated for interest on late payment, it does not give the contractor the right to assume that he can treat the late payment as an involuntary loan. Neither does it indicate that the subcontractor is giving up his right under the Housing Grants, Construction and Regeneration Act 1996 to suspend work, or his contractual right (if any) to determine his own employment, for nonpayment. Unfortunately, clause 3.2.2 does not impose any further subcontract terms on the contractor.

If the architect does decide to check the contractor's form of subcontract and it is not a recognised form, Figure 8.2 shows some of the matters for which it should make provision. Some of them are now imposed by the Act. It is essential that the subcontract steps-down the relevant main contract provisions. The architect should remember that he has no responsibility for the subcontract and if he checks it, he will do so merely to ensure that the contractor has avoided situations which might have repercussions on the main

Figure 8.2
Subcontract checklist

- Subcontractor's obligations
- Information: supply and timing of documents, drawings and details; custody and confidentiality; design liability
- Instructions, variations and valuations
- Access to site for subcontractor; access to subcontract works for architect; regulations
- Non-exclusive possession of the site
- Assignment and subcontracting
- Vesting of property and insurance
- Commencement, completion and extension of the subcontract period
- Completion date and making good of defects
- Payment in instalments*
- Withholding of money*
- Suspension of obligations*
- Financial claims
- Damages for delay
- Determination
- Finance (No 2) Act 1975
- Disputes procedure – adjudication,* arbitration

* Clauses imposed by the Housing Grants Construction and Regulation Act 1996.

contract. Any dispute between contractor and subcontractor is potentially disruptive.

8.2.3 Nominated subcontractors

MW 80 makes no provision for nominated subcontractors. Their use for works for which this form would be suitable is not envisaged. Having said that, there are various devices which can be used to provide for 'nominated' subcontractors if the architect or the employer are absolutely sure that they must 'nominate' a particular firm.

The widespread use of nominations in building contracts has come to be associated with certain laziness or shortage of time on the part of the architect. It often seems quicker and less trouble in the short term to put a sum in the bills of quantities or specification than to properly specify requirements at the outset. It is obviously a temporary expedient only and at some future date, usually sooner rather than later, the architect is faced with specifying the work in question and engaging in the complicated process of nomination. Difficult legal and administrative problems can result during the life of the contract – as numerous legal cases testify. The message is clear. Wherever possible, nomination should be avoided. If nomination is required consideration should be given to using JCT 80, which has detailed (but complex) provisions, or IFC 84, which provides for 'naming' subcontractors in a clause of equal complexity and some obscurity, or ACA 2, which has relatively simple 'naming' procedures. If 'nomination' is desired using MW 80, it is possible to use one of the following methods:

- Name one or a choice from several named firms in the specification
- Name a firm in an instruction directing the expenditure of a provisional sum in accordance with clause 3.7
- Include an appropriately worded clause in the contract
- Provide for the specialist firm to be directly employed by the employer.

There are severe pitfalls inherent in the use of any of these methods and the architect should seek the advice of an experienced construction contract consultant before he proceeds. Among the points to be considered are the following.

Naming in the specification

It is perfectly possible for the architect to name one firm in the specification which the contractor must use for carrying out a specific part of the work. If he does this, he places a considerable amount of power in the hands of such a firm who can, in effect, quote to the contractor whatever price it likes, secure in the knowledge what whichever contractor is awarded the contract, the specialist firm will be incorporated. If, as is not unknown, the specialist firm goes into liquidation before work begins on site, the architect has an instant problem with which to start his contractual duties.

The biggest difficulty with this type of nomination is that there is no provision in the contract to deal with the consequences. The precise extent of those consequences can be envisaged by glancing through clause 35 of JCT 80. The answer, of course, is to incorporate a fairly substantial clause in the contract to deal with the situation. Such a clause would require very careful drafting.

The alternative, whereby the contractor is given a choice of three or four names in the specification, is not strictly nomination at all. However, it does give a degree of control over the firms to be used, while allowing the contractor to seek competitive quotations. A subcontractor appointed by the contractor under this method would be the contractor's entire responsibility to the extent that, if it failed, it would be the contractor's job to find an alternative. Although the worst consequences of nomination can be avoided by this system, it is advisable to include a suitably worded clause in the contract.

Naming by an instruction in respect of a provisional sum

If it is intended to operate this system, it is essential that the contractor knows what is intended at the time of tender so that he can make suitable provision in his price. A big danger is that the contractor might strongly object to the firm in question when it is named for the first time in an instruction. There is no machinery to deal with his objection and the efficient progress of the work can be seriously affected. All the comments regarding the naming of a single firm in the specification are equally applicable to this case and an amendment to clause 3.2 and 3.7 is indicated.

Including a clause in the contract

The trouble with this is that once substantial clauses are added dealing with particular circumstances, in this case nomination, there is the risk of the contract being considered the employer's 'standard written terms of business' which brings it within the scope of the Unfair Contract Terms Act 1977. There is no doubt that a suitable clause can be formulated and, if nomination is what is wanted, it is probably the safest way to achieve it. Nomination can proceed along time-honoured procedures and potential difficulties can be provided for. On the other hand, to add a nomination clause like clause 35 in JCT 80 would effectively double the size of the contract – to say nothing of the ancillary documentation which would be necessary.

The employer directly employing the specialist firm

This may seem an attractive solution, but there are dangers, not least the problem of integration with the works (see section 8.4).

8.3   Statutory authorities

MW 80 does not expressly mention statutory authorities. However, clause 5.1 states that the contractor must comply with any statute, any statutory instrument, rule or order or any regulation or byelaw applicable to the works. This means, for example, that he must not contravene the Planning Acts or regulations made in pursuance thereof and he must comply with the Building Regulations and, of course, the Construction (Design and Management) Regulations 1994. By necessary implication, therefore, he must allow statutory undertakers to enter the site and carry out work which they alone are empowered to do.

Certain crucial parts of virtually all contracts are carried out by statutory undertakers such as local authorities, gas, water and electricity suppliers. When they carry out work solely as a result of their statutory rights or obligations, they are in a special category quite separate from subcontractors or employers' directly employed firms. If completion of the works is delayed as a result of a supplier carrying out or failing to carry out work in pursuance of its statutory duty, the contractor will be entitled to an extension of time in accordance with clause 2.2.

If the contractor employs a statutory undertaking to carry out work which is not part of its statutory rights or duties, the statutory undertaking ranks as an ordinary subcontractor. Delay caused by the undertaking in such a case would not entitle the contractor to an extension of time under MW 80. An example should make the principle clear.

It is usual to include a provisional sum in the specification to cover the cost of connecting the electrical system of a dwelling to the mains. The contractor may also, with the architect's written permission, sublet the electrical wiring and fittings in the dwelling to the electricity supplier. The mains connection is part of the supplier's statutory obligations; the internal wiring and fittings are not part of the supplier's statutory obligations, but a matter of contract between the contractor and the supplier. If the mains connection delays the completion date, the contractor is entitled to an extension of time. If the internal wiring and fittings delay the completion date, the contractor is not entitled to any extension of time.

Statutory undertakers have no contractual liability when they are carrying out their statutory duties – a sore point with many architects – but in certain cases they have a liability in tort. Outside their statutory duties, they are in the same position as anyone else if they enter into a contract to carry out work.

The contractor must also give all notices required by statute etc. and must pay all fees and charges in respect of the works, provided that they are legally recoverable from him, but not otherwise. He is not entitled to be reimbursed and he is deemed to have included the necessary amounts in his price. The exception to this, of course, is if the charge is a necessary result of an architect's instruction. In such a case, the amount of the charge will form part of the valuation of the instruction carried out under the provisions of clause 3.6.

Clause 5.1 states that the contractor is not liable to the employer under the contract if the works do not comply with statutory requirements provided that:

- He has carried out the works in accordance with the contract documents or any of the architect's instructions *and*
- If he has found a divergence between the contract documents or the architect's instructions and statutory requirements, he has immediately given him a written notice specifying the divergence.

The contractor is not liable if he fails to find a divergence which actually exists: *London Borough of Merton* v. *Stanley Hugh Leach Ltd* (1985) 32 BLR 51. Although the contractor may be freed from liability to the employer, his duty to comply with statutory requirements remains. Thus, the local authority may serve notice on him if work, built correctly in accordance with the contract documents, does not comply with the Building Regulations. In such a case, the architect would have to act speedily to issue appropriate instructions if the employer was to avoid a substantial claim at common law for damages.

The contractor is not entitled to take any emergency measures to comply with statutory requirements even though delay might cost the employer money. His obligation to give the architect immediate written notification remains. If the emergency concerns part of the structure which is actually dangerous, the contractor has a responsibility under the general law to take whatever measures are necessary to make the structure safe. If is difficult to see how he could make any valid claim in such circumstances.

Clause 5 has been augmented by JCT Amendment MW9 to make provision for compliance of the parties with the CDM Regulations. The idea is to make compliance with the regulations a contractual duty so that breach of the regulations is also a breach of contract. Clause 5.6 has been inserted to provide that the employer 'shall ensure' that the planning supervisor carries out all his duties under the regulations and that, where the principal contractor is not the contractor, he will also carry out his duties in accordance with the regulations. There are also provisions that the contractor, if he is the principal contractor, will comply with the regulations (clauses 5.7, 5.8 and 5.9).

Every architect's instruction potentially carries a health and safety implication which should be addressed. The CDM Regulations pose a formidable list of duties on the planning supervisor. Some of these duties must be carried out before work is started on site. If necessary actions delay the issue of an architect's instruction or once issued delay its execution, the contractor will be entitled to extension of time and, depending on circumstances, he may have a common law claim for damages for breach of an express term of the contract. These are matters about which all parties should take great care.

8.4 Works not forming part of the contract

The contract makes no provision for the employer to enter into a contract with anyone other than the contractor to carry out any part of the work on site while the contract work is being carried out. It is quite common for the employer to wish to engage others to do certain work or, indeed,use some of his own employees. The reason may be because the employer has a special relationship with the firm or individual, for example in the case of a sculptor, artist or landscaper, or because the employer wants complete control over a particular operation. When using MW 80, direct engagement by the employer can be achieved in one of two ways:

(1) With the consent of the contractor
(2) By including a special clause to that effect.

It is never a good idea to bring third parties onto a site during the contract period. Three dangers which merit special consideration are:

(1) It is easy for the contractor to claim that such persons have disrupted his work and/or delayed the completion date. They are very difficult claims to refute because introducing third parties clearly does not help the contractor and can usually be seen to cause at least some degree of hindrance. The contractor may be able to claim damages at common law and an extension of time under clause 2.2.

(2) The contractor may acquire grounds to determine his employment under the contract. Clause 7.3.1.2 provides for the contractor to determine if the employer 'or any person for whom he is responsible ... interferes with or obstructs the carrying out of the Works ...'. The matter is dealt with in more detail in section 12.3.2.

(3) For insurance purposes, persons directly employed by the employer are deemed to be persons 'for whom the Employer is responsible' (clause 6.1). They are not deemed to be subcontractors. Therefore the employer may have uninsured liabilities. The directly employed persons may well have their own insurance cover, but it is the architect's duty to advise the employer to obtain the necessary cover through his own broker. The cover should be for the employer and those persons for whom he is responsible in respect of acts or defaults occuring during the course of the work. It is a complex business and best left in the broker's hands.

Statutory undertakers acting outside the confines of their statutory duties may be considered to be directly employed by the employer if they are not paid by the contractor and under his control. It is always prudent to ensure that the contractor is responsible for all the work to be carried out during the currency of a contract. It removes some possible areas of dispute and promotes efficiency. If the employer insists on having directly employed persons, remember that, for work to be considered as not forming part of the contract, it must:

- Be the subject of a separate contract between the employer and the person who is to provide the work *and*
- Be paid for by the employer direct to the person employed, not through the contractor.

8.5   *Summary*

Assignment

- A different concept from subcontracting
- Neither party may assign without the other's consent
- Under the general law, either party may assign benefits, but not duties
- It may be beneficial to amend clause 3.1.

Sub-contracting

- The contractor may sublet with the architect's consent
- There is no contractual relationship between the employer and the subcontractor
- The subcontract terms may have important repercussions for the employer.

Nominated sub-contractors

- No provision in MW 80
- If nominated subcontractors are required, a different contract form should be considered
- Devices can be used to enable nominations to be made when using MW 80, but there are pitfalls.

Statutory authorities

- The contractor must allow them to enter the site
- He must comply with statutory requirements
- In pursuance of their statutory duties, they are not liable in contract, but may provide grounds for an extension of time

- Not in pursuance of their statutory duties, they are liable in contract and may be subcontractors or persons for whom the employer is responsible
- The contractor is not liable to the employer if he works to the contract documents, provided that he notifies the architect of any divergence he finds
- There is no provision for emergency work
- Compliance with the CDM Regulations is a contractual requirement
- Extension of time and other claims may arise from compliance with the CDM Regulations.

Work not forming part of the contract

- MW 80 makes no provision for directly employed persons
- The contractor must consent or a special clause must be included
- Fertile ground for claims
- Possible ground for determination
- Insurance implications
- Should be avoided.

CHAPTER NINE
POSSESSION, COMPLETION AND DEFECTS LIABILITY

9.1 Possession

9.1.1 Introduction

If there is no express term in the building contract, a term will always be implied that the contractor must have possession in sufficient time to allow him to finish the works by the contract completion date: *Freeman* v. *Hensler* (1900) 64 JP 260. To have possession of something is the next best thing to ownership. If the owner of this book lends it to a friend, the friend can defend his claim to it against anyone except the original owner. The same principle applies to a builder in exclusive possession of a site. He is in control of the site and has the power to refuse access to anyone else, including the employer. In practice, this stern rule is modified by the operation of numerous statutory regulations, allowing entry by the representatives of various statutory bodies, and by the express and implied terms of the contract.

In legal terms, the contractor is said to have a licence from the owner of the site to occupy it for the period of time necessary to carry out and complete the works. The period of time is the period stated in the contract or any extended period. During the contractor's lawful occupation the employer has no power under the general law to revoke the licence, but the contract may contain express terms giving such power, for example, in the case of lawful determination of the contractor's employment. Case law suggests that the absence of such an express term may pose awkward problems if the contractor refuses to give up possession: *Hounslow Borough Council* v. *Twickenham Garden Developments* (1971) 7 BLR 81. In general, however, if the contractor retains possession of the site after the contract period or any extended period has expired, he is in the position of a trespasser and can be removed under the general law.

This contract does contain an express term requiring the contractor to give up possession immediately if the employer determines the contractor's employment under the provisions of clause 7.2. Although there is no similar specific requirement when the contractor determines his own employment under clause 7.3, a term to that effect will be implied.

There is no express provision for access to the works for the architect equivalent to clause 11 of JCT 80, but such a term must be implied to allow the architect and his authorised representatives to carry out their duties under the contract.

9.1.2 Date for possession

JCT 80, IFC 84, ACA 2 and GC/Works/1 have an Appendix, Time Schedule or Abstract of Particulars in which important dates, sums of money and other variables are to be filled in by the parties before the contract is signed. The Appendix becomes a vital contract document. MW 80 has no such Appendix, but important dates etc. are to be inserted in appropriate places in the clauses themselves. This procedure should not pose problems in view of the brevity of the form.

Clause 2.1 relates to 'Commencement and completion', but it does not make reference to a date for possession. It does, however, leave a space for a date to be inserted on which the works 'may be commenced'. It will be implied that this date is the latest date on which the employer must give possession of the site to the contractor. The inclusion of the word 'may' could be significant. The straightforward interpretation of the clause appears to be that the date to be inserted is the earliest on which the contractor will be allowed to commence carrying out the works, but not necessarily the latest. In effect, it appears that the contractor would be within this clause if he started work in the fourth month of a six month contract, provided that he finished by the contract completion date. Contrast with JCT 80, clause 23.1, where the contractor 'shall thereupon begin the Works ...'. Of course, in practice, the contractor could well be in danger of having his employment determined in accordance with clause 7.2.1:

> 'if the Contractor without reasonable cause fails to proceed diligently with the Works or wholly suspends the carrying out of the Works before completion.'

In *Greater London Council* v. *Cleveland Bridge & Engineering Co Ltd* (1986) 8 Con LR 30, the court was of the opinion that a requirement for the contractor to proceed with due diligence and expedition must be interpreted in the light of other requirements as to time. The contractor may, therefore, be able to argue that he is proceeding regularly and diligently provided he can meet the completion date and it is a moot point whether he could be said to be suspending something he had not yet begun. In any case, he might well argue that he had reasonable cause if, in fact, the contract period was very generous. 'Regularly and diligently' has now been comprehensively defined by the Court of Appeal: *West Faulkner* v. *London Borough of Newham* (1993) 9 Const LJ 232. (See also section 5.1.2.)

If the employer fails to give the contractor possession so that he can begin the works on the stated date, it will be a serious breach of contract. The contractor will have a claim for damages at common law and the completion date will cease to be operative: *Rapid Building Co Ltd* v. *Ealing Family Housing Association Ltd* (1985) 1 Con LR 1. In such circumstances the contractor's obligation would be to complete within a reasonable time. Although that does not mean that the contractor has unlimited time in which to carry out the works, it does mean that there is no date from which liquidated damages can run and, therefore, they are not deductible. If the employer's failure to give possession lasts more than a few days, it is our view that the contractor may well have grounds to consider the employer's breach as an intention to repudiate the contract. The contractor should serve notice on the employer (Figure 9.1). In any case, failure to give possession clearly requires negotiation between the employer and the contractor to achieve an amicable settlement and the architect should remind the employer of his duties before the due date (Figure 9.2) and give him his advice if the employer is in breach (Figure 9.3).

Although the architect's power to instruct the contractor certainly extends to postponing the work, postponement is not the same as failure to give possession. If the architect postpones the work, the contractor will have possession of the site, but the carrying out of the work will be suspended. The contractor may well wish to use the time to reorganise his site arrangements, repair site offices, improve his security, etc.

When the contractor has completed the work, he normally gives up possession, but he has a restricted licence to continue to enter the site to deal with such defects as are notified to him under clause 2.5, Defects liability (see section 9.3).

Figure 9.1
Letter from contractor to employer if possession not given on the due date

REGISTERED POST OR RECORDED DELIVERY

Dear Sir

PROJECT TITLE

Possession of the site should have been given to us on [*insert date*] to enable us to commence the works on [*repeat date*] in accordance with clause 2.1 of the conditions of contract. Possession was not given to us on the due date.

This is a serious breach of contract for which we will require appropriate compensation and we reserve all our rights and remedies in this matter. Without prejudice to the foregoing, we suggest that a meeting would be useful and look forward to hearing from you.

Yours faithfully

Figure 9.2
Letter from architect to employer before date for possession

Dear Sir

PROJECT TITLE

The contractor is entitled, by the terms of the contract, to take possession of the site on the [*insert date*].

Will you be certain that everything is ready so that the contractor can take possession? Failure to give possession on the due date is a serious breach of contract which cannot be remedied by a simple extension of the contract period; the contractor may be able to claim substantial damages or even treat the contract as repudiated.

Please let me know immediately if you anticipate any difficulties.

Yours faithfully

Figure 9.3
Letter from architect to employer if he fails to give possession on the due date

Dear Sir

PROJECT TITLE

I understand/have been notified by the contractor [*delete as appropriate*] that you were unable to give him possession of the site on the date stated in the contract as the date on which he may commence the works.

You will recall that in my letter of the [*insert date*] I pointed out that failure to give possession is a serious matter.

It is something which you must negotiate with the contractor if you wish to avoid the charge of repudiation and heavy damages. With your agreement, I will try to negotiate on your behalf but, since this is outside the terms of my appointment, I should be pleased to have your written authorisation to act for you in this way and your agreement to pay my additional fees and costs on a time basis as laid down in my original conditions of appointment.

Yours faithfully

9.2 Practical completion

9.2.1 Definition

Clause 2.1 states that the works must 'be completed by' a date to be inserted. The words have their ordinary meaning, that is to say the completion of the works must not take place after the stated date, but it may take place before the stated date. Note, however, that there is no provision for partial possession by the employer. This omission is entirely reasonable in view of the small-scale nature of the works for which this contract is intended to be used and the correspondingly short time scale. It is extremely unlikely that the employer will need to take possession of part of the works before completion. If it is decided that provision must be made for partial possession it is worth considering the use of IFC 84 Instead, incorporating the provision for partial possession detailed in Practice Note IN/1, page 10. If phased completion is to be provided, MW 80 is not suitable and either JCT 80 or IFC 84 with the Sectional Completion Supplement or ACA 2 would be appropriate. It is not envisaged that partial possession, much less sectional completion, would be a feature of work for which MW 80 was being considered.

Clause 2.4 states that the architect must 'certify the date' when in his opinion the works have reached practical completion and the contractor has complied sufficiently with clause 5.9. Clause 5.9, of course, requires the contractor to provide and to ensure that any subcontractor provides information for the preparation of the health and safety file required by the CDM Regulations. Therefore, the architect is to certify the date by which both criteria have been satisfied. Only one date may be certified and not, as some architects believe, one date for each criterion. In accordance with clause 1.2, the architect must issue the certificate in writing. Although no time scale is indicated, the architect must issue the certificate within a reasonable time of the certified date because it is a particularly important stage in the contract (see section 9.2.2). In practice, he should issue the certificate immediately.

Note that it is the architect's opinion which is required by the contract, not that of the employer or the contractor. Despite the significant consequences of the architect's certificate, 'practical completion' is nowhere defined in the contract. It does not mean substantially or almost complete and the precise meaning has exercised several judicial minds. On balance, it seems to mean the stage at which there are no defects apparent and only very trifling items remain outstanding: *H.W. Nevill (Sunblest) Ltd* v. *Wm Press &*

Son Ltd (1981) 20 BLR 78. What qualifies as 'trifling' will depend on circumstances. It is not thought that the architect is justified in withholding his certificate until every last screw and spot of paint is in place. That would indeed be completion, but the contract clearly intends something rather short of that by the use of the word 'practical'. Within these guidelines the architect is free to exercise his discretion. He would not be justified in issuing his certificate, in any event, if items remained to be finished which would seriously inconvenience the employer. Note that the architect cannot issue the certificate even when in his opinion practical completion has been achieved. The CDM provisions must also be satisfied. (See above and the effect on the deduction of liquidated damages in section 10.2.2.)

The architect is under no obligation to issue lists of outstanding items if he withholds his certificate. Clerks of works often consider it part of their duty to supply the contractor with so-called 'snagging lists'. It is a bad practice because the contractor tends to assume that when he has completed the lists, his obligations are at an end; therefore, disputes sometimes occur. The contractor's obligations should be clear from the contract documents and the duty to make sure that the work is complete in accordance with the contract lies with the contractor, not with the architect or the clerk of works (if employed). More particularly, any 'snagging lists' should be prepared by the contractor's person-in-charge as part of his normal supervision of the works. Whether he does or not, is not the architect's concern.

The architect's duty to issue a practical completion certificate does not depend on any request by the contractor. He must carry out his duty as soon as he is satisfied that the criteria have been satisfied. Many architects arrange a handover meeting to which the employer and any consultants are invited. It should be remembered, however, that the architect cannot transfer responsibility for certifying practical completion to the employer, although it may be a prudent move to see that he is happy with the building before he takes possession. In practice, it is much more likely that the employer will wish to take possession before the architect is thoroughly satisfied. In such circumstances the architect must strictly observe his duty and refuse to issue his certificate until he is so satisfied. The contractor will then gain no benefit and may be at some disadvantage in completing the work. He may complain to the employer who may, in turn, complain to the architect who should put his position to the employer in writing for the record (Figure 9.4). If the architect submits to pressure, he may well leave himself open to a future claim for negligence, not only from the

137

Figure 9.4
Letter from architect to employer if he has taken possession of the works before practical completion

---

Dear Sir

PROJECT TITLE

I refer to your letter of the [*insert date*].

I confirm that, in my opinion, practical completion has not been achieved. Therefore, it is my duty about which I have no discretion, to withhold my certificate. I note, however, that you have agreed with the contractor to take possession of the building. Although I think you are unwise, I respect your decision and I will continue to inspect until I feel able to issue my certificate. At that date, the defects liability period will commence. Naturally, the contractor has a very great interest in obtaining a certificate of practical completion and you must expect him to continue to complain until, in my opinion, practical completion is achieved.

Yours faithfully

---

employer but also from third parties to whom the architect may have given a warranty that he will properly carry out his duties.

9.2.2 Consequences of practical completion

The issue of the certificate of practical completion is of particular importance to the contractor because it marks the date when:

- The defects liability period commences (clause 2.5)
- The contractor's liability for frost damage ends (clause 2.5)
- The contractor's liability for liquidated damages ends (clause 2.3)
- The employer's right to deduct full retention ends and half the retention held becomes due for release within 14 days (clause 4.3)
- The machinery culminating in the issue of the final certificate is set in motion (clause 4.5)
- The contractor's liability to insure under clause 6.3A ends.

9.3 Defects liability period

9.3.1 Definition

The defects liability period is inserted for the benefit of both parties. It allows a period of time for defects to appear and to be corrected with the minimum of fuss. Any defect which is the fault of the contractor is a breach of contract on the part of the contractor and, without such a period, the employer would have no contractual remedy. He would be left to his common law rights. Moreover, if there were no defects liability period, the contractor would have no right to re-enter the site to remedy the defects. If the employer suffers some loss as a direct result of the defects and the mere remedying of those defects is not adequate restitution, it is always open to him to take action at common law to obtain damages from the contractor for the breach.

Contractors commonly hold two mistaken views about the defects liability period:

(1) That the contractor's liability for remedying defects ends at the end of the defects liability period
(2) That the contractor is liable to correct anything which is showing signs of distress or with which the employer is not satisfied.

With regard to the first, the contractor's liability does not end at the end of the defects liability period. What does end is his privilege to return and remedy his breach. After that time, the contractor remains liable, but after proper notice the employer can simply pursue an action at common law for damages, if he so wishes, without giving the contractor the opportunity to return.

The second mistaken view probably owes its origin to the practice of referring to the defects liability period as the 'maintenance period'. Architects and contractors alike are guilty in this respect (even GC/Works/1 and ACA 2 use the term) although maintenance implies a heavier responsibility than simply making good defects. Repolishing, cleaning and generally keeping a building in pristine condition could be said to be maintenance for which the contractor has no responsibility. The employer's dissatisfaction with the building is also of little consequence in itself (see section 9.3.2) if the contractor has carried out his obligations. If instructed to carry out what amounts to matters of routine maintenance, the contractor should write to the architect making the position clear (see Figure 9.5).

9.3.2 Defects, excessive shrinkages and other faults

Clause 2.5 requires the contractor to make good 'defects, excessive shrinkages and other faults'. Case law has established that the phrase 'other faults' must be interpreted *ejusdem generis* with defects and excessive shrinkages; that is to say, faults of the same kind. Read in this way 'other faults' appears to add little if anything to the contractor's liability. A defect can only mean something which is not in accordance with the contract. If the employer is unhappy about the paintwork, it could be that the contractor has not applied it correctly in accordance with the specifications. On the other hand, it could be that the specification is inadequate. Only the former situation would give rise to liability on the part of the contractor. An inadequate specification is usually the architect's responsibility.

It is clear that not all instances of shrinkage fall within the defects liability clause. Shrinkages have to be excessive. The intention appears to be to exclude those shrinkages which could be said to be an unavoidable consequence of building operations. This is an eminently sensible approach, but one which could result in dispute because what is excessive to the architect may be trifling to the contractor. Indeed, the whole question of shrinkages is fraught with difficulty. They are the contractor's liability only if they result from

Figure 9.5
Letter from contractor to architect regarding routine maintenance

Dear Sir

PROJECT TITLE

We have received your instructions dated [*insert date*] which you purport to issue under the provisions of clause 2.5 of the conditions of contract. Among the items listed as defects for us to make good are [*list defects complained of*].

Clause 2.5 requires us to make good defects, excessive shrinkages or other faults provided that, among other things, they are due to materials and workmanship not in accordance with the contract. In this instance, the materials and workmanship are demonstrably in accordance with the contract. What you are in effect instructing us to do is to carry out items of routine maintenance. Such is not our obligation under the contract and you have no power to issue such an instruction.

Yours faithfully

workmanship or materials which are not in accordance with the contract. In practice, since the employer holds the purse-strings, it is the contractor who has to convince the architect that the shrinkage is not his responsibility. Shrinkage usually occurs due to loss of moisture or thermal movement. A common example is the shrinkage which occurs in timber after the building is heated. The architect will have specified a maximum moisture content for the timber, and subsequent shrinkage can only be because the timber was installed with, or allowed to develop, too high a moisture content, or the architect's specification was wrong. It is a question of fact rather than law, but note that it is no defence for the contractor to say that the architect's specified moisture content was impossible to achieve under normal site conditions. It is well known that it is difficult to maintain a low moisture content under the normally damp conditions which prevail on site, but that is not to say that such conditions cannot be improved by the use of suitable temporary heaters, ventilation, etc. The contractor's obligation is to provide workmanship and materials in accordance with the contract and he would have been well aware of the architect's requirements at tender stage. What he is really saying, therefore, is that he found it too expensive to comply with the architect's specification, and that is no defence at all.

9.3.3 Frost

The contractor's liability to make good frost damage is limited to damage caused by frost which occurred before practical completion. This is perfectly reasonable since he was in control of the works up to, but not after, practical completion. Damage due to frost occurring after practical completion is the responsibility of the employer. In practice, there should be no great difficulty in detecting the difference. Frost damage after practical completion may be due, for example, to faulty detailing, unsuitable materials or lack of proper care by the employer. Note that the test is not when the damage occurred, but when the frost, which resulted in such damage, occurred.

9.3.4 Procedure

The defects liability period starts on the date of practical completion as stated in the architect's certificate. If he does not insert any period of time in the contract, the period will be three months. There is really no good reason for limiting the period to three months because there is no connection between the length of the contract period and the defects liability period. Whether the contract period is long or short, there will be some items of work completed just before practical completion and it is these items which the architect must consider when advising the employer of the length of defects liability period required in a particular case. In general, there is probably much to be said for always inserting twelve months as the length of period on the basis that the building will be tested against all four seasons of the year. It is probably true that the contractor will include a slightly increased tender figure if twelve months is included as the defects liability period, but there is really no reason why he should do so. It probably stems from a mistaken idea of the limits of his liability (see section 9.3.1). The final certificate will, of course, be delayed, but only $2\frac{1}{2}\%$ of the retention will be outstanding and payment will then probably closely follow the architect's certificate that all defects have been made good (see Chapter 11).

The defects etc. which the contractor is to make good are those 'which appear within' the defects liability period. The wording suggests that any defects which have already appeared before practical completion could not be included as defects which the contractor must make good. In practice, the situation is not as bad as that. Defects which were apparent (sometimes referred to as 'patent defects') before practical completion would preclude the architect from issuing his certificate of practical completion (see section 9.2.1). Moreover, since no certificate is conclusive under this contract, the contractor's obligation to carry out the work in accordance with the contract is not reduced by the issue of any such certificate and if the contractor is so misguided as to refuse the opportunity of remedying the defectings during the defects liability period, the employer retains his common law rights intact. The danger is that if the architect certifies practical completion while there are some patent defects, half the retention will be released and the contractor may never return. The employer's common law rights will be useless if the contractor has gone into liquidation and the employer may, rightly, say that the architect was negligent in the issue of his certificate and look to him for recompense. If the architect does overlook some defects before issuing his certificate, the sensible thing to do is to include them with the defects which actually appear 'within' the period. This may not be precisely what the contract says, but it does no violence to the contractor's rights. This approach was noted with approval in *William Tomkinson* v. *Parochial Church Council of St Michael* (1990) 6 Const LJ 319.

Unlike the equivalent clause (17) in JCT 80, no procedure is laid down for the architect to issue a schedule of defects or even notify the contractor that defects exist. However, since the contractor cannot know of any defects which may appear within the period unless the architect does notify him, it is thought that the architect can use his powers under clause 3.5 to instruct the contractor to carry out urgent remedial action during the defects liability period and to issue an instruction at the end of the period listing all the defects which have appeared during the period. It is prudent to organise an inspection a few days before the period expires so that the architect can issue his final list of defects on the final day. The contractor should be prompt to respond if some items are not his responsibility (see Figure 9.6).

There is no time limit set for the contractor to make good the defects, but the architect is not obliged to certify that making good has been achieved until he is satisfied (to do otherwise would be negligent) and the issue of the final certificate (clause 4.5.1.1) is dependent upon the architect's certificate under clause 2.5. The contractor must carry out his obligations within a reasonable time. What is reasonable will depend on:

- The number and type of defects
- Any special arrangements to be made with the employer with regard to access.

If the contractor fails to carry out his obligations, the architect may put the compliance procedure under clause 3.5 into operation (see section 4.3). All defects are to be made good by the contractor entirely at his own cost 'unless the Architect shall otherwise instruct'. Some commentators have been greatly concerned by this last phrase, even going so far as to suggest that it can only mean that the architect can instruct the contractor to remedy defects at the employer's expense. With great respect, we suggest that such a singular view is nonsense. There is no doubt that the wording could be clearer, but it covers two distinct situations which might arise:

(1) If the defects are partly the fault of the contractor and partly contributed to by some default of the architect or the employer. In such circumstances it would be wrong to expect the contractor to remedy the defects entirely at his own expense and

Figure 9.6
Letter from contractor to architect after receipt of schedule of defects

Dear Sir

PROJECT TITLE

Thank you for your instruction number [*insert number*] dated [*insert date*] scheduling the defects you require making good now that the defects liability period has ended.

We have carried out a preliminary inspection and we are making arrangements to make good most of the items on your schedule. However, we do not consider that the following items are our responsibility for the following reasons:

[*list, giving reasons*]

We shall, of course, be happy to attend to such items if you will let us have your written agreement to pay us daywork rates for the work.

Yours faithfully

the architect might wish to direct him regarding the extent to which the contractor would be expected to bear the cost. The key words in the clause are 'entirely' and 'unless'.

(2)   If the employer does not wish the contractor to remedy certain defects because, for example, to do so would seriously disrupt the employer's business. In such circumstances, the architect might wish to instruct the contractor not to make good such defects.

In either of the above situations, the architect must be sure to discuss the matter thoroughly with the employer before he issues any instruction. In the second case, the architect must obtain a letter from the employer authorising him to instruct the contractor that making good is not required (Figure 9.7). If the architect instructs, note that there is no provision for him to authorise any deduction from the contract sum.

Although clause 2.5 does also empower the architect (as other commentators suggest) to instruct the contractor to make good defects at the employer's expense, it is not something which the architect should ever consider doing. When making good has been completed, the architect must issue a certificate to that effect. The certificate has important implications with regard to the issue of the final certificate.

9.4   Summary

Possession

- Must be given to the contractor in sufficient time for him to complete the works
- Cannot be revoked by the employer under the general law, but the contract may give such power
- Is the same as the date on which works may be commenced in MW 80
- Is the contractor's right, and failure on the part of the employer is a serious breach of contract
- Ends at practical completion.

Practical completion

- Is a matter for the architect's opinion alone subject to case law.

Figure 9.7
Letter from architect to employer if some defects are not to be made good

---

Dear Sir

PROJECT TITLE

I understand that you do not require the contractor to make good the following defects:

[*list*].

These defects are included in my schedule of defects issued at the end of the defects liability period. In order that I may issue the appropriate instructions in accordance with clause 2.5, I should be pleased if you would confirm the following:

(1) You do not require the contractor to carry out making good to the defects listed in this letter.
(2) You waive any rights you may have against any persons in regard to the items listed as defects in the above-mentioned schedule of defects and not made good.
(3) You agree to indemnify me against any claims made by third parties in respect of such defects.

Yours faithfully

- Requires the architect to issue a certificate; there is no provision for sectional completion or partial possession
- Bestows important benefits on the contractor.

Defects liability period

- Benefits the contractor
- The end of the period does not mean the end of the contractor's liability for defects
- Defects are severely limited in scope
- May be of any length
- Is not a maintenance period
- Only those defects appearing within the period are covered under the contract
- Defects must be notified to the contractor during or at the end of the period
- Defects must be made good at the contractor's own cost
- The architect has power to instruct the contractor not to make good some defects or to instruct what proportion of cost must be borne by the contractor; check with employer first
- When making good is complete, the architect must issue a certificate to that effect.

CHAPTER TEN
CLAIMS

10.1 General

It is traditional in the construction industry for claims by the contractor for both extra time and extra money to be linked together. That pattern is followed in this chapter, but it should be borne in mind that there is no necessary link between time and money. The grant of an extension of time is not a pre-condition to a monetary claim for 'loss and/or expense' under any of the JCT Forms, neither does an extension of time automatically entitle the contractor to 'loss and/or expense'. There can be money claims for both prolongation and disruption, and claims give rise to hard feelings. MW 80 is unique among the JCT contracts in that it contains no clause entitling the contractor to 'direct loss and/or expense' except for the provision in clause 3.6 requiring the valuation of variations to include 'any direct loss and/or expense incurred by the contractor due to regular progress of the works being affected by compliance with' a variation instruction or 'due to compliance or non-compliance by the employer with clause 5.6' (obligations in regard to the planning supervisor and the principal contractor).

But the absence of such a clause should not mislead either employer or architect into thinking that the contractor cannot make financial claims. He can; but such claims must be pursued at common law by way of arbitration or litigation and must be based on breach of some express or implied term of the contract or some other legal wrong. The architect cannot deal with such claims, unless the employer expressly authorises him to do so, and the contractor agrees. However, the architect may be the cause of such claims if, for example, he is late in issuing the contractor with necessary information or instructions, or if he fails to carry out his duties under the contract and the contractor suffers loss as a result.

The case of *Croudace Ltd* v. *London Borough of Lambeth* (1986) 6 Con LR 72 establishes this point. In carrying out his duties under the contract the architect owes a duty to the employer to act fairly (*London Borough of Merton* v. *Stanley Hugh Leach Ltd* (1985) 32 BLR 51) as he does when certifying: *Sutcliffe* v. *Thackrah* [1974] 1 All ER 319. If the architect fails to carry out his duties properly, this is a breach of contract for which the employer is liable in damages. If that happened, no doubt the employer would seek to recover from the architect.

Although the architect owes a duty to the employer to carry out his functions under the contract in a professional manner with proper care and skill, he does not, it seems, owe a duty of care in tort direct to the contractor. This certainly appears to be the case where there is an arbitration agreement (as in MW 80) which enables the architect's certificates and opinions to be revised: *Pacific Associates Inc* v. *Baxter* (1988) 16 Con LR 90, where an earlier (and some people think) more realistic decision of a very experienced official referee to the contrary was doubted. In the earlier case (*Michael Sallis & Co Ltd* v. *Calil and William F. Newman & Associates* (1987) 12 Con LR 68) it had been held that an architect owed a duty of care to the contractor to act fairly as between him and the employer in matters such as the issue of certificates and the grant of extensions of time and that the contractor might recover damages direct from an unfair architect. Recent decisions, however, may be signalling a new era of architects' liability to the contractor if reliance can be established: *Henderson* v. *Merritt Syndicates* (1994) 69 BLR 29. (See section 4.1.)

10.2   Extension of time

10.2.1   Legal principles

At common law, the contractor is bound to complete the work by the date for completion stated in the contract, unless he is prevented from doing so by the employer's fault or breaches of contract (*Percy Bilton Ltd* v. *Greater London Council* (1982) 20 BLR 1) and the employer's liability extends to the architect's wrongful acts or defaults within the scope of his authority. In the absence of an extension of time clause, neither the employer nor the architect would have any power to extend the contract period. Clause 2.2 deals with extension of the contract period and is linked with clause 2.3 which provides for liquidated damages. Both those provisions were revised in October 1988. Clause 2.2 was amended again in 1992 and 1994. It cannot be over-emphasised that the only effect of the architect granting an extension of time is to fix a new date for completion and, incidentally, to relieve the contractor from paying liquidated damages at the stated rate for the period in question.

10.2.2 Liquidated damages

There are many misconceptions about liquidated damages. The most important point is that the sum stated as such is recoverable whether or not the employer can prove that he has suffered any loss as a result of late completion or even if, in the event, he suffers no loss at all. The object of the liquidated damages clause is to fix an amount which is a pre-estimate of the quantum of damages which the employer may suffer through late completion and if the contractor is late in completing it is irrelevant to consider whether in fact there is any loss: *BFI Group of Companies Ltd* v. *DCB Integrated Systems Ltd* (1987) CILL 348, a case on MW 80 holding that liquidated damages are payable even if there is no loss.

Although often referred to by contractors as 'the penalty clause', a penalty is not enforceable. A sum is treated as liquidated damages (and so recoverable) if it is a fixed and agreed sum which is no more than reasonable and is a genuine pre-estimate of the loss likely to be incurred, or a lesser sum, estimated at the time the contract is made. It matters not that the estimate is a poor one in fact. Deciding whether a sum is liquidated damages or a penalty can be difficult where the parties have inserted a complex set of provisions. However, the courts have indicated that, in deciding the issue, they will not take account of hypothetical situations. They are much more likely to take a pragmatic approach: *Phillips Hong Kong Ltd* v. *The Attorney General of Hong Kong* (1993) 9 Const LJ 202.

Under MW 80 the amount inserted as liquidated damages is usually relatively small in relation to the potential loss to the employer from late completion, but it is bad practice to pluck a figure out of the air. The architect in consultation with the employer should have calculated the amount of liquidated damages carefully at pre-tender stage.

To set the figure at the right level, the architect should discuss it carefully with his client and them make a calculation. This is sometimes difficult. In the case of profit-earning assets, there is no problem. All that then need be done is to analyse the likely losses and additional costs. The following should be considered:

- Loss of profit on a new building, e.g. rental income, retail profit, etc.
- Additional supervision and administrative costs – including additional professional fees
- Any other financial results of the late completion, e.g. storage charges for furniture in the case of a domestic building.

An alternative method , but one more open to criticism, is to use one or other of the several formulae used in the public sector, such as that put forward by the Society of Chief Quantity Surveyors in Local Government, which gives an approximation to a detailed analysis of all individual costs, but this calculation should also be based on verifiable data. The figure thus arrived at must be inserted in clause 2.3. If no figure was inserted, no liquidated damages would be payable. The same result would follow if no completion date was inserted in clause 2.1 since there must be a date from which liquidated damages can run: *Kemp* v. *Rose* (1858) 1 Giff 258. However, liquidated damages are exhaustive of the employer's remedies for late completion and in *Temloc Ltd* v. *Erill Properties Ltd* (1987) 39 BLR 30 the inclusion of '£NIL' in a JCT 80 appendix entry was held to preclude any claim for damages.

Clause 2.3 makes it clear that liquidated damages are payable at the specified rate only if the works are not completed by the original completion date or extended contract completion date. They are payable by the contractor at the stated rate per week for the period between the stated completion date and the date of practical completion as certified by the architect under clause 2.4. The clause confers an express right on the employer to deduct liquidated damages from monies due to the contractor, e.g. progress payments, or he may recover them as a debt.

The certificate of practical completion no longer signifies the date on which the works have reached that stage. In addition, it certifies that the contractor has sufficiently complied with the CDM Regulations (where applicable) (see section 9.2.1). So, by the time the certificate is issued, the works may have been at practical completion stage for several weeks. The operative date for the end of liquidated damages, therefore, may be the actual date of completion, because clause 2.3 states: 'If the Works are not completed ...'. If damages are to be deducted, the appropriate notices under clauses 4.4.2 or 4.5.1.3, as appropriate, must be given. Strangely, an additional notice must be given not later than the issue date of the final certificate if the contractor wishes to deduct from the final certified sum.

10.2.3   Extending the contract period

Clause 2.3 was redrafted in October 1988 and, unlike the long and complicated provisions in other JCT contracts, it is short and sweet. It empowers the architect to grant an extension of time for completion:

'if it becomes apparent that the Works will not be completed ... *for reasons beyond the control of the Contractor*, including compliance with any instruction of the Architect ... whose issue is not due to a default of the Contractor ...' (emphasis added)

It is not clear whether the italicised phrase in fact extends to delay which is the fault of the employer. Sensibly, it ought to do so, but the case law suggests otherwise and this is reinforced by the express reference to architect's instructions. In *Wells* v. *Army & Navy Co-operative Society Ltd* (1902) 86 LT 764, a very similar phrase, *'other causes beyond the contractor's control'* was not to extend to delays caused by the employer or his architect. The last sentence of clause 2.2 was added in the wake of *Scott Lithgow* v. *Secretary of State for Defence* (1989) 45 BLR 1, when the court surprisingly held that, in some circumstances, subcontractors may not be under the control of the contractor. So far as MW 80 is concerned, it is now clear that subcontractors and suppliers are within the control of the contractor.

In *Peak Construction (Liverpool) Ltd* v. *McKinney Foundations Ltd* (1970) 1 BLR 111, the Court of Appeal ruled that liquidated damages clauses and extension of time clauses were both to be interpreted strictly *contra proferentem* against the employer and should not be read to mean that an employer can recover damages for delay for which he was partly to blame. It was said that:

'if the employer is in any way responsible for the failure to achieve the completion date, he can recover no liquidated damages at all and is left to prove such general damages as he may have suffered.'

The effect of this is that, although the architect must protect the employer's right to deduct liquidated damages by making extensions of time if appropriate, the architect has no power to make extensions of time for any reasons which do not fall within the grounds set out in the extension of time clause.

That case concerned an in-house form of contract, but its authority was upheld by the Court of Appeal in *Rapid Building Co Ltd* v. *Ealing Family Housing Association Ltd* (1985) 1 Con LR 1, a case which involved a JCT 63 contract.

Figure 10.1 illustrates the contractor's duties in claiming an extension of time under clause 2.3.

Figure 10.2 sets out the architect's duties in relation to such a claim. The procedure under clause 2.3 is straightforward:

The contractor must notify the architect in writing 'if it becomes

Figure 10.1
Flowchart of contractor's duties in claiming an extension of time

```
                        ┌──────────┐
                        │  START   │◄──────────────┐
                        └──────────┘               │
                             │                     │
                             ▼                     │
                      ╱─────────────╲              │
                     ╱   Apparent    ╲             │
                    ╱  that works will ╲    Yes     │
                    ╲  be completed    ╱────────────┤
                     ╲  by due date   ╱             │
                      ╲─────────────╱               │
                             │ No                   │
                             ▼                      │
                   ┌───────────────────┐            │
                   │  Examine reasons  │            │
                   └───────────────────┘            │
                             │                      │
                             ▼                      │
                      ╱─────────────╲               │
                     ╱    Due to     ╲              │
                    ╱ reasons beyond  ╲    Yes      │
                    ╲  contractor's   ╱─────────────┤
                     ╲   control     ╱              │
                      ╲─────────────╱               │
                             │ No                   │
                             ▼                      │
                      ╱─────────────╲               │
            No       ╱    Due to     ╲              │
         ┌──────────╱   architect's   ╲             │
         │          ╲  instructions   ╱             │
         │           ╲───────────────╱              │
         │                  │ Yes                   │
         │                  ▼◄──────────────────────┘
         │       ┌───────────────────────┐
         │       │ Notify architect in    │
         │       │      writing           │
         │       └───────────────────────┘
         │                  │
         │                  ▼
         │       ┌───────────────────────┐
         │       │ No requirement to provide│
         │       │ further information or  │
         │       │ use best endeavours to  │
         │       │ minimise delay, but     │
         │       │ clearly advisable       │
         │       └───────────────────────┘
         │                  │
         │                  ▼
         │              ┌────────┐
         └─────────────►│  STOP  │
                        └────────┘
```

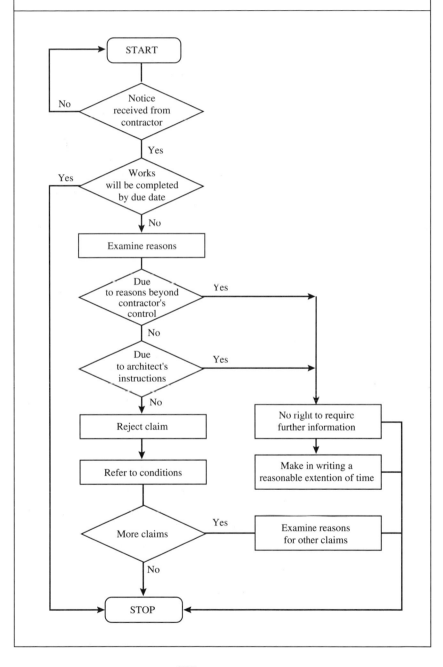

Figure 10.2
Flowchart of architect's duties in relation to claim for extension of time

apparent' that the current completion date will not be met. The contract refers to the fact that 'the Works will not be completed by the date for completion inserted in clause 2.1 hereof (or any later date fixed . . .)' under clause 2.2. There is no reference to delays to progress.

The architect must then make, in writing, a *reasonable* extension of time. The contract does not say when the architect must do this, but we suggest that he must grant the extension of time as soon as possible. Certainly he should not spike the application and must grant the extension before the current date for completion is passed, if practicable. The architect must of course be satisfied that the completion date will not be met because of 'reasons beyond the control of the contractor'. The recent case of *John Barker* v. *Portman Hotels* (1996) CILL 1152 contains much useful guidance for the architect in arriving at the proper extension of time. It is not sufficient for the architect to form an impression of the period; the period must be calculated. The judge referred with approval to the contractor's approach to the problem – putting the programme on a computer and inputting delays.

Failure properly to grant an extension of time may result in the contract completion date becoming at large, and liquidated damages becoming unenforceable.

Figure 10.3 is a suggested letter awarding an extension.

What happens if the contractor fails to notify the architect that the completion date will not be met? Clearly he is in breach of contract in not doing so because the obligation rests on him, and it is suggested that the contractor's failure to give notice is a matter which may be taken into account by the architect in determining the extension of time: *London Borough of Merton* v. *Stanley Hugh Leach Ltd* (1985) 32 BLR 51. It is certain that the contractor's notice is not a precondition ('condition precedent') to the grant of an extension of time and it is up to the architect to monitor progress and in an appropriate case award an extension of time even if the contractor has not notified him. The primary purpose of an extension of time clause is to protect the employer against the loss of liquidated damages following some act of hindrance or prevention.

10.3 Money claims

10.3.1 General

Although there is limited scope in clause 3.6 for the architect to include direct loss and expense in a valuation if it is incurred by the

---

Figure 10.3
Letter from architect to contractor granting
extension of time under clause 2.2

Dear Sirs

PROJECT TITLE

On [*insert date*] you notified me that the works would not be completed by [*insert clause 2.1 date or extended date*] because of [*specify reasons, e.g. the national strike of building trades employees in pursuit of a pay claim*].

I accept that this falls within clause 2.2 of the contract, and in accordance therewith I hereby grant you an extension of time of [*specify period*]. The revised date for completion is now [*insert date*].

Yours faithfully

---

contractor as a direct result of regular progress being affected by compliance with a variation instruction or compliance or non-compliance by the employer with his duties in respect of the planning supervisor or principal contractor (if other than the main contractor), there is no clause in the contract which entitles the contractor to make financial claims against the employer whether arising from prolongation or disruption. This leads some people to suppose that this is a risk which the contractor must price. Nothing could be further from the truth. The absence of a 'direct loss and/or expense' clause merely means:

- There is no contractual provision for ascertaining and paying money claims
- The architect has no power to quantify or agree such claims.

The contractor must pursue his claims in arbitration or litigation, unless he and the employer can agree the amount. The absence of a 'claims' clause is a grave defect of this form of contract and not a benefit as some employers appear to think. The problem can, of course, be overcome by including a suitably worded clause as an amendment to the contract. There are plenty of precedents, for example, IFC 84, clauses 4.11 and 4.12.

10.3.2   Types of claims

Financial claims are commonly referred to in the industry as *ex-contractual claims*. These are claims made outside the express pro-visions of the contract, usually as a result of breach of its terms, express or implied. They are also called common law claims, and increasingly are based on *implied terms* relating to co-operation and non-interference by the employer.

The situation is well-illustrated by *Holland Hannen & Cubitt (Northern) Ltd* v. *Welsh Health Technical Services Organisation* (1981) 18 BLR 80 where it was pleaded on behalf of contractors under a JCT 63 contract that there were implied terms whereby the employer contracted that he and his architect:

- Would do all things necessary on their part to enable [the con-tractors] to carry out and complete the works expeditiously, economically and in accordance with the contract.

Conversely, that neither the employer nor his architect:

- Would in any way hinder or prevent [the contractors] from carrying out and completing the works expeditiously, economically and in accordance with the contract.

We suggest that such terms are to be implied when the contract is in MW 80 form and this being so opens up a wide area for claims, which may also arise in tort.

If claims of this nature are made by the contractor they must be made against the employer. He may well seek the architect's advice, and possibly authorise him to deal with them. But the contractor is not entitled to reimbursement because he is losing money; it is up to him to establish that the employer or the architect is in breach. Valid claims can only arise because the contractor suffers loss through the fault of the employer or those for whom he is responsible in law.

Table 10.1 summarises the MW 80 clauses which may give rise to claims. Useful books for reference purposes are:

- *Powell-Smith and Sims' Building Contract Claims*, 3rd edition, by David Chappell, Blackwell Science, 1997.
- *The Presentation and Settlement of Contractors' Claims*, by Geoffrey Trickey, Spon, 1983.

10.4 Summary

Claims for time and money are distinct; there is no necessary connection between the two.

Liquidated damages

Liquidated damages are:

- A genuine pre-estimate of likely loss or a lesser sum
- Recoverable without proof or loss
- Recoverable only if the contractor has not completed the works by the original or extended completion date.

Extension of time

The architect is bound to grant a reasonable extension of time for completion if:

- The contractor notifies him that the works will not be completed by the current completion date because of reasons beyond his

Table 10.1
Clauses that may give rise to claims under MW 80

Clause	Event	Type
1.2	Failure to issue further necessary information	CL
2.1	Failure to allow commencement on the due date by lack of possession or otherwise	CL
2.2	Failure to give extension adequately or in good time	CL
2.3	Wrongful deduction of damages	CL
2.4	Failure to certify practical completion at the proper time	CL
2.5	Wrongful inclusion of work not being defects etc. Failure to issue certificate that the contractor has discharged his obligations	CL CL
3.1	Assignment without consent	CL
3.2.1	Unreasonably withholding consent to subletting	CL
3.4	Unreasonably or vexatiously instructing removal of employees from works	CL
3.5	Failure to confirm oral instructions Instructions altering the whole character or scope of the work Wrongful employment of others to do work	CL CL CL
3.6	Variations Compliance or non-compliance of the employer with clause 5.6 Wrongful omission of work to be done by others	C C CL
4.1	Errors or inconsistencies in the contract documents	C
4.2	Failure to certify payment if requested Failure to certify payment for materials properly on site	CL CL
4.3	Failure to certify payment at practical completion	CL
4.5	Failure to issue final certificate	CL
4.6	Contribution, levy and tax changes	C

Table 10.1 *Contd*		
Clause	**Event**	**Type**
5.1	Divergence between statutory requirements and contract documents or architect's instruction	C
7.2.1	Invalid determination Payment after determination	CL C
7.2.2	Invalid determination	CL
7.3.3	Payment after determination	C

Key
C = Contractual claims   CL = Common law claims
Contractual claims are usually dealt with by the architect.
Common law claims are usually dealt with by the employer.
Note: There is no loss and/or expense clause in this contract. Such claims have to be made at common law.

control, including an architect's instruction not occasioned by his default.

If the architect fails to grant an extension of time for completion, the contract time may become 'at large' and liquidated damages will be irrecoverable.

Money claims

- There is limited contractual provision for the contractor to be reimbursed 'direct loss and/or expense'.
- Any claims by the contractor must be pursued under the general law
- Such claims can only arise because the contractor suffers loss through the architect's fault or that of the employer.

CHAPTER ELEVEN
PAYMENT

11.1 Contract sum

The contract sum is the sum of money which is inserted in Article 2 of the Articles of Agreement. It is stated to be exclusive of VAT which means that VAT payments which may be necessary will be additional to this sum (clause 5.2). How far this will affect the employer will depend on his status, from the point of view of being able to reclaim VAT, and the work involved in the contract.

The figure in Article 2 will be the contractor's tender figure or such figure as the parties agree, perhaps after negotiation. The importance of the sum cannot be over-emphasised. It is the sum for which the contractor has agreed to carry out the whole of the works as shown on the contract documents. MW 80 is a 'lump sum contract' which means that the contractor is entitled to payment provided he completes substantially the whole of the works. In theory, the existence of a system of interim payments does not alter the position and if the contractor abandons the work before completion, the employer is entitled to pay nothing more. Once written into the contract, the contract sum may be adjusted only in accordance with the contract provisions (Table 11.1). Errors or omissions in the computation of the contract sum are deemed to be accepted by employer and contractor. Inconsistencies may be corrected in accordance with clause 4.1 (see section 3.1.2). It is quite possible for the contractor to make a considerable error in his calculations to such an extent that the contract becomes no longer viable from his point of view. If the error is undetected before the contract is entered into, there is nothing the contractor can do about the situation, except perhaps submit an *ex-gratia* (on grounds of hardship) claim, with little hope of success. This situation must be avoided if possible because it is unsatisfactory from all points of view. The employer may indeed think that he is gaining, but a contractor in this position has very little incentive to work efficiently and every reason to submit claims at every opportunity. Unfortunately, unless the contract documentation includes quantified

Table 11.1
Adjustment of contract sum under MW 80

Clause	Adjustment
2.5	Defects etc. during the defects liability period
3.6	Variations
3.7	Provisional sums
4.1	Inconsistencies in the contract document
4.5	Computation of the final amount
4.6	Contribution, levy and tax changes
6.3A	Insurance money
6.3B	Making good of loss or damage

schedules or bills of quantities, it is very difficult to check the contractor's pricing. Some errors may be obvious, but where the contract documents consist of drawings and specification, the contractor's pricing strategy may be obscure unless he separately submits a detailed breakdown of the figure.

11.2 *Payment before practical completion*

Clause 4.2.1 sets out the procedure for what is termed 'progress payments'. The architect must certify 'progress payments to the contractor' at not less than four-weekly intervals. The phrase is ambiguous. It may mean that he is to certify *progress payments to the contractor*, such certificates being sent, as is now customary, to the employer with a copy to the contractor; or it may mean that he is to certify *progress payments* and send the certificates to the contractor. In the last edition we suggested that the latter interpretation is correct since there is no other indication to whom the certificate should be sent. Since then, we have heard some persuasive arguments for the contrary view which causes us to reassess our opinion. In order to give the contract business efficacy, the architect must issue his certificates to the employer, such certificates being certificates of *progress payments to the contractor*. This view is reinforced by reference, later in the clause, to the employer paying 'the amount so certified'. Clearly, he cannot pay these amounts unless he knows what they are, and to adopt the latter construction requires the contractor to present the certificate for payment or for the employer to pay on a copy certificate received from the architect. To imply either of these procedures is unrealistic. The employer must receive the certificate, while the contractor receives a copy.

The employer must pay within 14 days of the date of issue of the certificate. In practice, it is usual for the contractor to send four-weekly statements of account for the architect's consideration.

Although there is no provision for any other system of payment, it may be more convenient, on small works, to agree a system of stage payments. If it is desired to operate in this way, it is important to make the necessary amendments to the printed conditions and to ensure that the contractor is aware of the change at tender stage. It will have a considerable impact on his pricing strategy.

The amount to be included in the certificate is to consist of:

- The value of works properly executed
- Amounts ascertained or agreed under clause 3.6 (variations) or clause 3.7 (provisional sums)

- The value of any materials and goods which have been reasonably and properly brought on the site for the purpose of the works and which are adequately stored and protected against the weather and other casualties
- Amounts calculated under the provisions of clause 4.6.

Less

- A retention of 5%
- The total amount due in previous certificates.

Each of the above items requires careful consideration, as follows.

The value of works properly executed

There has been a difference of opinion about the meaning of the word 'value' in this context. The generally accepted view is that it is the valuation obtained by means of reference to the priced document in relation to the amount of work actually carried out. Defective work is not included and a retention percentage is deducted. The retention is intended to deal with problems which might arise. The biggest problem would be the contractor's going into liquidation immediately following a payment. The alternative view of the meaning of 'value' derives from this possibility. From the employer's point of view, the value of the contractor's work is the value of the whole contract less the cost of completing the work with the aid of another contractor and additional professional fees. The additional cost would be considerable and incapable of being met from the retention fund.

The latter view finds little favour with contractors since the certificates issued at any given stage would bear no relation to the money expended by the contractor. Indeed, during the early part of a job, the certificates might even show a minus figure. The two views might be termed contractor value and employer value. The services of a quantity surveyor would be necessary to determine the probable cost of completion at the time of each certification. His task would not be easy. It is possible to operate this system if it is made clear to the contractor at the time of tender so that he can take extra financing charges into account. Case law suggests that the courts will favour the traditional contractor value view of 'value'.

'Properly executed' refers to the fact that the architect must not include the value of work which is defective, that is, not in accordance with the contract. If the architect does certify defective work

166

because, perhaps, the defect does not make itself immediately apparent, the correct procedure is to omit the value from the next certificate. This should pose no difficulties unless, of course, the contractor abandons the work first.

Amounts ascertained or agreed under clauses 3.6 and 3.7

This refers to any variations which are valued before the date of the certificate and any instructions which the architect may issue with regard to provisional sums which result in an adjustment to be valued under clause 3.6.

The value of materials and goods etc.

The reasoning behind the inclusion of payments for unfixed materials on site is clear. It is to enable the contractor to recover, at the earliest possible time, money which he has already laid out. The architect need not include any materials which he considers have been delivered to the site unreasonably early for the sole purpose of obtaining payment, but in its present form the clause contains serious dangers for the employer. There is no provision for the contractor to provide proof of ownership before payment. There- fore, if unfixed materials are included in a certificate and the con- tractor does not own them, the employer could be faced with the prospect of paying twice for the same materials if the contractor goes into liquidation and the true owner claims the goods from site: *Dawber Williamson Roofing & Co Ltd* v. *Humberside County Council* (1979) 14 BLR 70. The prevalence of retention of title clauses in the supply contracts of builders' merchants makes this a very real danger. Such a retention of title clause cannot be overcome by anything which may be written into the contract. Many architects amend the clause by deleting the whole of the provision 'and the value of any materials or goods ... protected against the weather and other casualties'. Since there is no provision for payment for off- site materials, this will then mean that the architect does not have to certify any unfixed materials whatsoever.

A retention of 5%

This is covered in detail in section 11.7.

The total amount due in previous certificates

We are pleased to see that following our criticisms of the wording of this part of the clause in the last edition, JCT have changed it.

Previously it referred to previous payments made by the employer – something of which the architect would have no formal knowledge.

The certificate must state to what the payment relates in general terms and it must show the basis of calculation. To comply with the Housing Grants, Construction and Regeneration Act 1996, the employer must give the contractor a written notice, not later than five days after the issue of a certificate. The written notice must specify the amount of payment to be made. It probably does not matter if the employer forgets to give the notice provided that he pays the amount certified and pays it within the stipulated 14 days (see clause 4.4.1).

11.3   Penultimate certificate

The contract provides for a special payment to be made at practical completion (clause 4.3). This clause is titled 'penultimate certificate' because, although it is not expressly stated, it is clear that regular certificates will cease at this point because there is no more work to value. Indeed the only other payment to be made will be the final payment.

The penultimate certificate must be issued within 14 days of the date of practical completion as certified by the architect. The employer has 14 days, as before, in which to pay. The effect of the certificate is to release to the contractor the balance of the monies due on the full value of the works, plus half the retention already held. The employer retains only $2\frac{1}{2}\%$ of the total value. Obviously, it may not be possible for the architect to know the full value precisely at this stage, but he is to certify it 'so far as that amount is ascertainable at the date of practical completion'. In other words, he must do the best he can pending final computation. Amounts ascertained or agreed under clauses 3.6 and 3.7 are to be included and, again, reference is made to deducting the amounts of previous progress payments. The comments made earlier are also applicable to this clause, including the written notice of payment to be given by the employer.

11.4   Final certificate

The contract lays down a precise time sequence for the events leading up to, and the issue of, the final certificate (see the contract time chart, Figure 11.1).

The contractor's duty is to send the architect all the documenta-

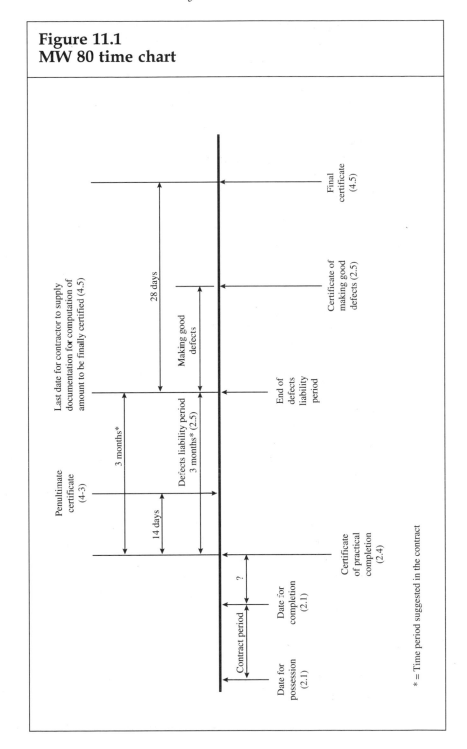

Figure 11.1
MW 80 time chart

tion he reasonably requires in order to compute the amount to be finally certified. The architect is entitled to request any particular supporting evidence he requires. The contractor has three months, or such other period as inserted in the contract, from the date of practical completion to send the information to the architect. The architect must issue the final certificate within 28 days of the receipt of the contractor's information. There is a proviso. The architect may not issue the final certificate until after he has issued the certificate of making good defects under clause 2.5. In practice, this will often be the deciding factor. If the contractor is late in sending the information to the architect, he is technically in breach, but it is of little consequence. He is simply delaying the time when he receives payment because the architect's 28 days does not begin to run until he receives the contractor's documentation.

The final certificate must state the amount remaining due to the contractor or, more unusually, the amount due to the employer. The latter situation will only occur if the architect has overcertified. The employer has 14 days, as before, in which to pay from the date of the final certificate. Clause 4.5.1.2 stipulates that the employer must give written notice to the contractor specifying the amount of payment to be made. The notice must be given not later than five days after the issue date of the final certificate. This is a similar provision to the notice provision for interim certificates and it is for the same purpose.

There is no provision in the contract for the contractor to agree the computations before the final certificate is issued, but it is customary to attempt to obtain his agreement. If the contractor delays sending his agreement to the architect this does not affect the architect's duty to issue the final certificate within 28 days. A letter from the architect to the contractor should make this clear (Figure 11.2).

11.5 *Effect of certificate*

It is refreshing to note that no certificate is stated to be conclusive. Thus, the issue of the certificate of making good defects does not preclude the architect from requiring the contractor to remedy anything which is not in accordance with the contract. Similarly, if the architect includes defective work in a four-weekly progress certificate, he can remedy the situation with the next certificate issued.

The contractor's liability is not reduced in any way by the issue of the final certificate which is not even conclusive as far as the com-

---

Figure 11.2
Letter from architect to contractor requesting
agreement to the computation of the final sum

Dear Sir

PROJECT TITLE

I enclose copies of the computation of the final sum to be certified.

I should be pleased if you would signify your agreement to the sum and the way in which it has been calculated by signing and dating one copy of the calculation in the space provided and returning it to me by the [*insert date*] at the latest.

If you have any queries, please telephone me as soon as possible, but you should note that, in any event, I have a duty under clause 4.5 of the Conditions of Contract to issue my final certificate no later than [*insert date*].

Yours faithfully

Copy: Employer

putations are concerned. *Crown Estates Commissioners* v. *John Mowlem* (1994) 70 BLR 1 considerably broadened the conclusive effect of the final certificate issued under JCT 80. It is thought that this case has no relevance to MW 80.

11.6   Withholding payment

If the employer fails to pay any amount due to the contractor by the final date for payment (i.e. 14 days from date of issue of a certificate), clause 4.2.2 in respect of interim and the penultimate certificate, and clause 4.5.2 in respect of the final certificate, require the employer to pay simple interest on the outstanding amount at the rate of 5% over the current Base Rate of the Bank of England.

If the employer believes himself to be entitled to withhold or deduct any amount from a payment to the contractor, he must give a written notice to the contractor not later than five days before the final date for payment. The notice must state the grounds for withholding payment and the amount of money withheld in respect of each ground. The information must be detailed enough to enable the contractor to understand the reason why he is not receiving the amount withheld. The contractor may, of course, seek immediate adjudication under the provisions of article 4.1 (see Chapter 13).

11.7   Retention

The reference to retention in this contract is remarkably brief. It is mentioned only in clauses 4.2.1 and 4.3 and then only in respect of the amount. In particular there is no reference to:

- The purposes for which the retention can be used
- Its status as trust money
- Keeping it in a separate bank account.

The reason for this is clearly to maintain the brevity of the contract as a whole, but there may be repercussions:

Use of retention

Clause 3.5 allows the employer to deduct the cost of employing others to carry out instructions with which the contractor has failed to comply, from monies due or to become due to the contractor.

Retention monies clearly fall within this clause, but the employer will, more commonly, deduct such monies from future certificates. If the contractor goes into liquidation before the work is complete, there is no doubt that the employer would be entitled to make use of retention monies to complete the work. There is no express provision enabling the employer to make use of the retention for any other purpose.

For example, the employer may not deduct the cost of taking out insurance if the contractor defaults. In practice, as mentioned earlier, the situation may suggest taking a broader view.

Trust money

Unlike the corresponding provisions of JCT 80 and IFC 84, the retention is not expressly stated to be trust money and it is not considered that any term would be implied from the general law to create such a trust.

Whether the retention is or is not a trust is of vital importance to the contractor for two reasons. First, a trust is governed by statute and, despite what the contract provisions may state, it is possible that the employer always has a duty to invest and to return an interest to the contractor. Where the retention is not stated to be a trust, the contractor is not entitled to any interest.

Second, where retention is stated to be trust money, the employer is in the position of trustee; he is merely holding the money for the contractor's ultimate benefit. It is, in no sense, the employer's money. Therefore, if the employer were to become insolvent, the trust money would not form part of his assets in the hands of his trustee in bankruptcy or liquidator. The contractor would be able to recover the full amount. If the money is not a trust, as in the case of MW 80, the contractor would have no better chance of recovery then any other unsecured creditor. MW 80, therefore, leaves the contractor at a severe disadvantage.

Separate bank account

Where the retention is trust money, it is often required to be placed by the employer in a separate bank account specially opened for the purpose. The result is that, in the event of the employer becoming insolvent, there is no difficulty in identifying the money. It is also good practice for any trust money to be dealt with separately from the employer's own money so that the investment income can be easily ascertained. Even if the contract is silent, it is now established law that the contractor can demand that trust money be kept in a separate bank account. The employer has a duty to put trust monies in a separate bank account whether or not so requested. Certainly, the employer may not mix trust money with his working capital: *Wates Construction (London) Ltd* v. *Franthom Property Ltd* (1991) 53 BLR 23. In the present case, since the retention is not stated to be trust money, the contractor cannot so demand.

11.8   Variations

Variations are covered by the extremely short clause 3.6. It gives the architect power to order:

- An addition to the works
- An omission from the works
- A change in the works
- A change in the order in which they are to be carried out
- A change in the period in which they are to be carried out.

It should be noted that the contractor has no right to object to any change in sequence or timing of the works.

This seems rather harsh, but taken in the context of the size and type of contract likely to be carried out under this form it is unusual to find problems arising in practice. In any case, the contractor is entitled to be paid for changing the sequence or timing, so he should not, in theory, suffer. The clause provides that the architect is to value all the types of variation listed. If a quantity surveyor is employed, the wise architect will leave such matters to him, but the contract is silent about the role of the quantity surveyor (see section 7.1) and he has no powers. Even if the architect does ask the quantity surveyor to undertake the valuation of variations, the architect bears the ultimate responsibility.

The contract allows the architect to agree a price with the contractor for any variation, but agreement must be reached before the contractor carries out the instruction. In practice, it allows the architect to invite and accept the contractor's quotation. On small works this is the sensible way of operating the provision if the additional or omitted work is anything other than more, or less, of the same.

If the architect does not agree a price with the contractor, the clause 3.6 lays down how the valuation is to be carried out. It is to be done on a 'fair and reasonable basis'. Although it is pleasant to think that 'fair and reasonable' is an objective concept, in practice it is the architect's opinion as to what is fair and reasonable that matters. When carrying out the valuation, he must use 'where relevant' prices in the priced documents. They are:

- The priced specification *or*
- The priced schedules *or*
- The contractor's own schedule of rates.

It is a matter for the architect to decide whether the prices are relevant. It is considered that unless the work being valued is precisely the same, carried out under the same conditions, he is at liberty to ignore the prices in the price document. This is because even a slight change in the conditions under which work is being carried out will have a marked effect on the cost to the contractor. The status of the contractor's schedule of rates has already been discussed (Chapter 3). If the architect waits until the first variation has to be valued before asking the contractor to provide it, the way is open for the contractor to 'massage' the rates. Whether or not the contractor does so, the architect will retain his suspicions. If the schedule is not provided before the job starts on site, any later provision will be suspect and the architect will have an unenviable task trying to value variations properly.

The valuation must include 'any direct loss and/or expense incurred by the Contractor due to regular progress of the works being affected by compliance with' the variation instruction. The purpose of this provision is to reimburse the contractor for the loss and/or expense directly resulting from the variation, but not forming part of the cost of the varied work itself. In other words, it covers the consequential effect of the introduction of the variation upon other unvaried work. It is suggested that it also covers other items of cost not directly related to the variation, which cannot be covered by prices in the priced documents.

There is no express requirement for the contractor to submit vouchers or other information to assist the architect in arriving at a fair and reasonable valuation, but it would be a foolish contractor who refused reasonable requests in this respect. If the contractor refuses to supply information, the architect must carry out the valuation to the best of his ability and the contractor will have no valid claim if he receives less than he expects. The valuation will, of course, include an element for profit, overheads and so on, as usual.

Clearly a fair and reasonable valuation must also include the valuation of work or conditions not expressly covered in the instruction but affected by that instruction. For example, if the architect issues an instruction to change the doors from painted plywood faced to natural hardwood veneered and varnished, it might well affect the sequence in which the contractor hangs the doors, the degree of protection required and the difficulty of painting surrounding woodwork. All this must be taken into account in the valuation; it is 'direct loss and/or expense' associated with the variation, as is other disruption or prolongation directly flowing from the variation.

11.9 Provisional sums

Clause 3.7 empowers the architect to issue instructions to the contractor directing him how provisional sums are to be expended. Provisional sums are usually included when the precise cost or extent of work is not known at the time of tender. The purpose is to have a sum of money to cover the cost. When the architect issues his instruction, he must omit the sum and value the instruction in accordance with the principles in clause 3.6 (explained in section 11.8 above). It is possible to use clause 3.7 to nominate a subcontractor, but this is not a sensible course (see section 8.2.3).

11.10 Fluctuations

Clauses 4.6 and 4.7 deal with fluctuations. Clause 4.6 is used if it is intended that the bare minimum fluctuation to deal with contribution, levy and tax changes are to be allowed. Detailed provisions are to be found in Part A of the Supplementary Memorandum. In the case of short-term projects for which this form is intended to be used, this clause will be deleted, as the footnote makes clear.

Clause 4.7 states that, save for the limited provision of clause 4.6 if retained, the contract is to be considered fixed price. The contractor must carry the risk of increase in cost of labour, materials, plant and other resources. Of course, in the unlikely event of a general fall in costs, the contractor would gain.

11.11 Summary

Contract sum

- Exclusive of VAT
- The amount for which the contractor agrees to carry out the whole of the work

Payment

- MW 80 is a 'lump sum' contract
- May be adjusted only in accordance with the contract provisions
- Errors are deemed to be accepted by both parties.

Payment before practical completion

- The architect is to certify payments at not less than four-weekly intervals
- An alternative system of payment may be agreed between the parties
- The employer must pay within 14 days of the date of the certificate
- The employer must give five days notice of the amount to be paid
- The architect must decide the meaning of 'value' and inform the contractor at tender stage
- The employer may reserve his retention on the whole of the certified sum
- There are dangers in certifying unfixed materials.

Penultimate certificate

- The certificate must be issued within 14 days of the date of practical completion
- The employer must give five days notice of the amount to be paid
- Half the retention must be released.

Final certificate

- The contractor must send all the documents the architect requires to compute the final sum
- The contractor has three months from the date of practical completion to send them
- The final certificate must be issued within 28 days of the receipt of the contractor's information provided that the architect has issued his certificate of making good defects
- The employer must give five days notice of the amount to be paid
- The contractor does not have to agree the architect's computations.

Effect of certificates

- No certificate is conclusive in any respect
- The contractor's liability is not reduced by the final certificate.

Withholding payment

- Notice must be given five days before the payment is finally due
- 5% simple interest is payable on amounts not properly paid.

Retention

- The employer may deduct money from the retention fund to pay the cost of employing others in accordance with clause 3.5
- The retention is not trust money and need not be kept in a separate bank account
- The contractor has no better claim on the retention money than any other unsecured creditor if the employer becomes insolvent.

Variations

- The contractor cannot object to a change in the sequence or timing of the works
- The architect must check all valuations
- The architect may agree a price with the contractor before he carries out an instruction
- Otherwise valuations must be on a fair and reasonable basis using prices in the priced document if relevant
- There is no provision for the valuation of claims for loss and/or expense
- Provisional sum work must be valued in accordance with clause 3.6.

CHAPTER TWELVE
DETERMINATION

12.1 *General*

Under the general law, a serious breach by one of the parties to a contract may entitle the other (innocent) party to treat the contract as at an end. But it is not every breach of contract which entitles the innocent party to behave in this way. The breach must be 'repudiatory', i.e. it must be conduct which makes it plain that the innocent party will not perform his obligations, or else it must consist of misperformance which goes to the root of the contract.

Until all the facts have been investigated – and this must take place after the event – it is sometimes difficult to say at what point a breach has occurred which is sufficiently serious to entitle the innocent party to terminate the contract.

In common with most standard building contracts, MW 80 provides for either party to determine the contractor's employment under the contract by going through a prescribed procedure. The determination clause (7.0) attempts to improve on the common law rights of the parties; indeed, the clause goes on to specify what the rights of the parties are after there has been a valid determination. This clause was comprehensively amended in 1994. Among other things, it introduced the facility to serve notice by 'actual' delivery as well as by registered post or recorded delivery (clause 7.1). There is a deeming provision that notice by post is deemed to have been received 48 hours after the date of posting, excluding Saturday, Sunday and public holidays. Care should be taken with this provision, because it can be overturned by proof that delivery occurred on another day.

Determination of the contractor's employment is a serious step, and the consequences of a wrongful determination are serious. It has been held, however, that where a party honestly relies on a contract provision, although mistaken, the party has not repudiated the contract: *Woodar Investment Development Ltd* v. *Wimpey Construction UK Ltd* [1980] 1 All ER 571. It is best avoided if possible, and the process is fraught with pitfalls for the unwary. Quite apart from the possibility of an action for breach of contract by one party or the other if things go wrong, the employer is always placed in an impossible position as far as getting the project completed is concerned.

Even if the employer is successful in recovering his costs from the contractor, he can never recover the time which he has lost. In practice, too, the contractor is in an invidious position, whether he determines his own employment or the employer does so. The formalities laid down in the determination provisions must be followed exactly if a costly dispute is to be avoided.

The architect bears a heavy burden since he will have to advise the employer about his rights and the procedure to be followed, and the determination clauses are only too easy to misunderstand. A contractor who believes that he has grounds on which to determine his own employment under clause 7.0 should never attempt to do so without seeking competent legal advice.

Either the employer or the contractor may exercise the right to determine the contractor's employment and MW 80 deals separately with determination by the employer and by the contractor.

12.2 Determination by the employer

12.2.1 Grounds and procedure

The procedure for determination is set out in Figure 12.1 while the grounds which may give rise to determination and the procedure to be followed are specified in clause 7.2.

There are four separate grounds for determination in that clause. They are that the contractor:

(1) Fails to proceed diligently with the works without reasonable cause (clause 7.2.1) *or*
(2) Wholly or substantially suspends the carrying out of the works, before practical completion without reasonable cause (clause 7.2.1) *or*
(3) Fails to comply with the CDM Regulations
(4) Becomes insolvent in one of the ways specified in clause 7.2.2.

The first three of these grounds is described as a 'default'.

If the employer, on the architect's advice, decides to determine the contractor's employment, the architect must ensure that the procedure is followed precisely. There is provision for preliminary notice before the right of determination is exercised. The architect

Figure 12.1
Flowchart of determination by employer

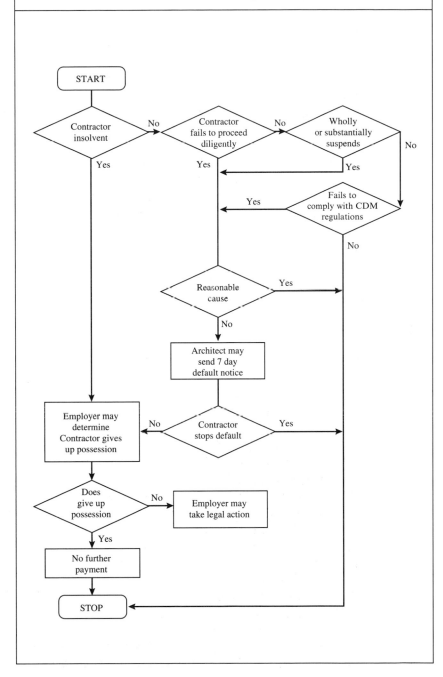

must give notice which specifies the default and which requires it to be ended. Figure 12.2 is the sort of letter which the architect might draft and it must be sent by registered post or recorded or actual delivery so that there is no doubt that the contractor has received it. This gives the contractor advance warning that the employer may exercise his right to determine. In many cases, that will be sufficient to stop the default immediately. 'Actual' delivery is a comparatively recent amendment. The architect must take care to obtain a receipt acknowledging delivery if this is the chosen method.

Before the architect advises the employer to determine the contractor's employment by notice he should consider the grounds carefully, as follows.

Fails to proceed diligently with the works

This is a breach of the contractor's basic obligation in clause 1.1 which requires him 'with due diligence and in a good and workmanlike manner' to carry out and complete the works in accordance with the contract documents. Whether or not the contractor is proceeding diligently is a factual question.

The case of *West Faulkner* v. *London Borough of Newham* (1993) 9 Const LJ 232 gives excellent guidance on the topic (see section 5.1.2).

Wholly or substantially suspends the carrying out of the works before completion

This ground was broadened in 1994 to include 'substantially suspending'. This overcomes the common problem that the contractor has not actually stopped working completely, but for all the good he is doing, he might as well have totally stopped. If the workforce dropped overnight from 40 operatives to just a couple of labourers, determination on this ground may be indicated. It is impossible to give precise guidance and each situation will depend on its own particular facts.

It should be noted that the contractor's action must be without reasonable cause – and a reasonable cause might well be, for example, the employer's failure to make progress payments under the architect's clause 4.2.1 certificates.

Fails to comply with the CDM Regulations

This is a new ground. It is thought the failure must be clear and unambiguous to justify a determination.

Figure 12.2
Letter from architect to contractor giving notice of default

REGISTERED POST, RECORDED OR ACTUAL DELIVERY

Dear Sir(s)

PROJECT TITLE

I hereby give notice under clause 7.2.1 of the contract that you are in default in the following respects:

[*Insert details of the default with dates if appropriate*]

If you continue the default for seven days after receipt of this notice, the employer may thereupon determine your employment under this contract without further notice.

Yours faithfully

Copies: Employer
          Quantity surveyor [*if appointed*]

The procedure for determining the contractor's employment is that if the contractor does not cease his default within 7 days of receipt of the default notice, the employer – not the architect – serves a notice by registered post or recorded or actual delivery, determining the contractor's employment. The notice operates from the date it is received by the contractor: *J.M. Hill & Sons Ltd* v. *London Borough of Camden* (1980) 18 BLR 31.

Financial failure of contractor

This is the fourth ground. Unlike some other contract forms, determination on the ground that the contractor is insolvent is not automatic. It is by notice. The specified grounds are if the contractor:

- Becomes bankrupt
- Makes a composition or arrangement with his creditors
- Makes a proposal in respect of his company for a voluntary arrangement for a composition of debts or a scheme of arrangement to be approved in accordance with the Companies Act 1985 or the Insolvency Act 1986
- Has a winding-up order made
- Passes a resolution for voluntary winding up (except for purposes of reconstructing the company)
- Has a provisional liquidator appointed
- Has an administrative receiver or administrator as defined under the Insolvency Act 1986 appointed.

These are all factual matters, and in some cases, e.g. where a receiver is appointed, it may be best for the architect to advise the employer not to determine the contractor's employment, since the receiver is bound to carry on with the company's contracts. In any event, specialist advice is indicated.

In the case of such insolvency, the contract provides that the employer need only give notice and determination takes effect on receipt of such notice by the contractor. There is no necessity for any warning notice such as is required in the case of a default.

The architect should draft a suitable letter for the employer's signature (Figure 12.3). Determination of the contractor's employment by the employer is governed by an important proviso. The right must not be exercised 'unreasonably or vexatiously'. In simple terms, this means that the procedure must not be instituted without sufficient grounds so as to cause annoyance or embarrassment. In

---

Figure 12.3
Letter from employer to contractor
determining employment

REGISTERED POST OR RECORDED OR ACTUAL DELIVERY

Dear Sirs

PROJECT TITLE

Despite the architect's letter to you dated [*insert date of architect's default letter*] you have failed to end your default[1].

In accordance with clause 7.2[2] of the conditions of contract, take this as notice that I hereby determine your employment under this contract without prejudice to any other rights or remedies which I may possess.

You are required to give up possession of the site of the works immediately, and should you fail to do so I shall instruct my solicitors to issue appropriate Court proceedings against you.

Yours faithfully

Copy: Architect

[1] Delete if determination due to insolvency.
[2] Substitute '7.2.2' if determination due to insolvency.

J.M. Hill & Sons Ltd v. *London Borough of Camden* (1980) 18 BLR 31 at p 49, the Court of Appeal took the view that 'unreasonably' in this context meant 'taking advantage of the other side in circumstances in which, from a business point of view, it would be totally unfair and almost smacking of sharp practice'. In *John Jarvis Ltd* v. *Rockdale Housing Association Ltd* (1986) 10 Con LR 51, the word 'unreasonably' in this context was said by the Court of Appeal to be a general term which can include anything which can be judged objectively to be unreasonable, while 'vexatious' connotes an ulterior motive to oppress or annoy.

12.2.2 Consequences of employer determination

Clause 7.2.3 sets what is to happen should the employer successfully determine the contractor's employment:

- The contractor must cease to occupy the site immediately. His licence to occupy it is at an end, and if he fails to do so, he becomes a trespasser in law. The contractor must remove himself, his men, plant, materials and tools from site.
- The employer is relieved of any obligation to make any further payment to the contractor until after completion of the works and the making good of any defects.
- The employer may recover from the contractor the additional cost of completing the works. Note that this is the additional cost to the employer and not the somewhat lesser cost to the contractor.
- The employer may recover from the contractor any expenses properly incurred by the employer as a result of the determination (e.g. the costs of re-tendering) and any direct loss and/or damage caused by the determination.

12.3 Determination by the contractor

12.3.1 General

The contractor has a right to determine his own employment for specified defaults by the employer and if he is successful in doing so, the results for the employer will be disastrous. Because of this, the architect should do everything in his power to prevent it from happening.

The practical consequences of a successful contractor determination are:

- The project will have to be completed by others, probably at a higher cost. In any event, professional and other fees will be involved, as well as other expenses.
- The completion date will be inevitably delayed.
- The employer may be faced with a liability to pay the contractor the profit he would have made if the contract had proceeded normally.

Figure 12.4 shows the procedure for determination by the contractor. The grounds and procedure for determination and its consequences are laid down in clause 7.3.

12.3.2 Grounds and procedure

There are six separate grounds which entitle the contractor to exercise his right to determine his employment under the contract. They are that the employer:

(1) Fails to discharge in accordance with the contract the amount properly due
(2) Interferes with or obstructs any certificate or the carrying out of the works
(3) Fails to make the premises available to the contractor at the right time
(4) Suspends the carrying out of the works for one month
(5) Fails to comply with the CDM Regulations
(6) Becomes insolvent in one of the ways specified in the clause.

It should be noted that there is no provision that these defaults must be 'without reasonable cause', and the only safeguard from the employer's point of view is that the contractor must not exercise his option to determine his employment 'unreasonably or vexatiously'.

As in the case of employer determination, the contractor must follow the procedure precisely as a wrongful determination may amount to a repudiation of the contract. When the contractor wishes to determine on one of the first five of the above grounds, he must first send a notice to the employer (not to the architect) by registered post or recorded or actual delivery giving notice of his intention to determine. That notice must specify the alleged default and require

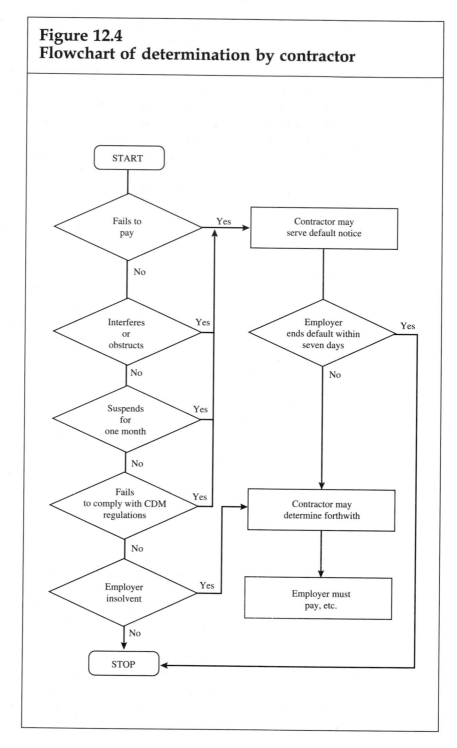

Figure 12.4
Flowchart of determination by contractor

that it ends (Figure 12.5). The employer has seven days from receipt of that notice in which to make good the default, and only if he continues the default for seven days may the contractor determine his employment. Determination is brought about by a further notice served on the employer by registered post or recorded or actual delivery and it takes effect on receipt (Figure 12.6).

In the case of the employer's financial failure (clause 7.2.4) no preliminary notice is required.

It is worthwhile considering the grounds for determination in detail because they have been put in to protect the contractor against common wrongdoings by employers.

The employer fails to discharge in accordance with the contract the amount properly due

This ground protects the contractor's right to be paid on time and expressly includes payment of VAT. Steady cash flow is as important to him as it is to the architect or his employer and in fact this provision is quite generous to the employer. Under clause 4.2.1 the employer has to pay to the contractor the amounts which the architect certifies as progress payments within 14 days of the date of the architect's certificate and payment is due to the contractor before that period expires.

If the employer does receive a notice under clause 7.3.1 he must pay at once, and if the architect knows about it, he should telephone the employer and advise him to pay immediately and confirm this by a letter along the lines of Figure 12.7.

Right from the outset, the architect must make sure that the employer understands the scheme of payments and the need to pay promptly on certificates. Where possible, financial certificates should be delivered by hand and a receipt obtained. If this is impracticable, they should be sent by registered post or recorded delivery. Once the 14 day period of grace has expired, of course, the contractor may refer to adjudication in any case, although most contractors prefer to rely on their option to determine employment.

If the contractor issues a writ and applies for summary judgment, on application by the employer, the matter will be stayed to arbitration.

The case of *C.M. Pillings & Co Ltd* v. *Kent Investments Ltd* (1985) 4 Con LR 1 has muddied the waters to some extent because it establishes that if the employer is genuinely disputing the amount of a certificate – and can produce hard evidence to back up his contention – the contractor's action may be unsuccessful in

Figure 12.5
Letter from contractor to employer giving notice of default before determination

REGISTERED POST OR RECORDED OR ACTUAL DELIVERY

Dear Sir

PROJECT TITLE

We hereby give you notice under clause 7.3.1.1/.2/.3 [*delete as appropriate*] of the conditions of contract that you are in default in the following respect:

[*insert details of the default with dates if appropriate*]

If you continue the default for 7 days after receipt of this notice, we may forthwith determine our employment under this contract without further notice.

Yours faithfully

Figure 12.6
Letter from contractor to employer determining employment after default notice

REGISTERED POST OR RECORDED OR ACTUAL DELIVERY

Dear Sir

PROJECT TITLE

We refer to the default notice sent to you on the [*insert date*].

Take this as notice that, in accordance with clause 7.3.1, we hereby determine our employment under this contract without prejudice to any other rights or remedies which we may possess.

We are making arrangements to remove all our temporary buildings, plant, etc. and materials from the works and we will write to you again within the next week regarding financial matters.

Yours faithfully

Figure 12.7
Letter from architect to employer advising immediate payment

Dear Sir

PROJECT TITLE

I refer to our telephone conversation today when I advised you to make immediate payment to the contractor of Certificate No 00 amounting to £[*insert amount*] in light of the service upon you of a notice preliminary to the contractor determining his employment under the contract.

You cannot assume that you have a full 14 days to pay because, technically, the certificate is not honoured until your cheque has been cleared through the bank. I suggest that you arrange to have your cheque delivered by hand to the contractor's office. If you allow the contractor to determine his employment the consequences will be considerable extra cost and delay to the project.

It is essential that you pay the certified progress payments within 14 days of the date on the certificate and it is up to you to make the necessary financial arrangements to ensure that funds are available at that time.

Yours faithfully

arbitration. But it is never sufficient for the employer to refuse to pay until he has satisfied himself that it is correct. The architect bears a heavy responsibility to ensure that it is correct.

It is also important to read the wording of the determination provision carefully. It does not say that the contractor may determine if the employer does not pay the amount certified, but rather refers to the 'amount properly due in respect of any certificate'. This wording would seem to leave open the employer's ordinary equitable and common law rights of set-off.

The employer or any person for whom he is responsible interferes with or obstructs any certificate or the carrying out of the works

This is a breach of contract at common law and interference or obstruction is a serious matter. In law, conduct of this kind is often referred to as 'acts of hindrance and prevention' and is a breach of an implied term of the building contract. The reference is to the carrying out of the works. The ground covers a wide range; instructing the architect not to issue a certificate of practical completion or countermanding the architect's instructions on site, for example, would fall within the clause. The architect will probably know from the contractor if this sort of thing is happening, and if the contractor does complain to the architect, and he thinks that the complaint is justified, he should write to the employer warning him of the likely consequences (Figure 12.8).

The employer or any person for whom he is responsible fails to make the premises available to the contractor

The use of the word 'premises' is curious as is the wording generally, although the intent is clear. It means the site. Clause 2.1 states quite simply that 'The Works may be commenced on [Date] and shall be completed by [Date]'. There is an implied term in this contract that the employer will give possession of the site to the contractor in sufficient time for him to complete his operations by the contractual date, and failure to do so is a serious breach of contract: *Freeman v. Hensler* (1900) 2 HBC 292.

The employer suspends the carrying out of the whole or substantially the whole of the works for a continuous period of at least one month

This ground is drafted very broadly and it is not the equivalent of, for example, JCT 80, clause 28 A, which refers to suspension by *force*

Figure 12.8
Letter from architect to employer warning of likely consequences of interference

---

Dear Sir

PROJECT TITLE

When I visited the site today, the contractor complained to me that [*insert details as appropriate, e.g. your office manager insisted that the contractor's men should not proceed with the renovation works to the Board Room*].

The contractor intimated to me that he was minded to determine his employment under clause 7.3.1.2 of the contract on the grounds that this conduct by your office manager amounts to interference with or obstruction of the carrying out of the works. He has previously complained in similar terms.

If the contractor does determine his employment, the consequences will be serious in terms of both time and cost. Determination is something to be avoided at all costs, and I suggest that you advise your staff that any instructions of this nature could have serious repercussions for you.

Yours faithfully

---

majeure, specified perils, civil commotion, etc. It is not known why this ground was inserted by the draftsman, because the contract does not expressly confer on the employer any right to suspend the works – and to do so would be a breach of contract – and even if the architect has the right to order a suspension under his general powers under clause 2.5, it would not fall under this clause. Any suspension must be *continuous* for a period of a calendar month.

Even though the employer has no power to suspend the carrying out of the work, he may do so, e.g. because he is in financial difficulties. The contractor could then exercise the right of determination under the clause if that suspension lasted for a month. But he would be better off to rely on his common law rights!

Fails to comply with CDM Regulations

This is a new ground and it is thought that the failure must be clear and unambiguous to warrant determination.

Employer's financial failure

These grounds (clause 7.3.2) parallel those of contractor insolvency in clause 7.2.2 (see Section 12.2.1) and no preliminary notice is required from the contractor. Theoretically, this ground is applicable even in the case of a local authority employer, although the whole clause is drafted on the assumption that the employer is an individual or a limited company since, technically, none of the events referred to is applicable in law to local authorities. This is not to say that local authorities cannot get into financial difficulties or become 'insolvent'.

12.3.3 Consequences of contractor determination

If the contractor determines his employment correctly, he is placed squarely in the driving seat. He must prepare an account which must set out:

- The value of work properly executed and materials properly on site for the works. The value must be ascertained as if the contractor's employment had not been determined, together with any other amounts due but not included.
- The cost of removal of all temporary buildings, plant, tools and equipment from site.

- Any direct loss and/or damage caused by the determination (this will include loss of profit and any other costs caused to the contractor).

The architect is not required to issue a certificate. The contract merely requires the employer to pay to the contractor the amount properly due after taking into account everything previously paid. The payment must be made within 28 days of submission of the contractor's account.

The clause stops short of requiring the employer to pay the balance of the amount on the contractor's account; hence the use of the phrase 'amount properly due'. It is envisaged that the employer will require his professional advisers to verify the contractor's account first. However, payment must be made within 28 days and it is no excuse for the employer to plead that he has not completed his checking process. It is suggested the 28 days will not begin to run until the account has been received in a form which, viewed objectively, will allow verification to take place.

Effect of determination on other rights

The contractor's right to determine his employment is expressly stated to be without prejudice to any other rights or remedies which he may possess. This means that his ordinary rights at common law are preserved, as are those of the employer. This makes it plain that, for example, the contractor can choose if he wishes to terminate the contract under the general law.

12.4 Summary

Determination by employer

The employer may determine the contractor's employment if:

- Without reasonable cause the contractor fails to proceed diligently
- Without reasonable cause the contractor substantially or wholly stops work
- The contractor fails to comply with the CDM Regulations
- The contractor becomes insolvent.

The employer may not exercise this right unreasonably or vexatiously. The procedure laid down must be followed exactly. The architect must advise the employer.

Determination by contractor

The contractor may determine his own employment if:

- The employer does not pay on time
- The employer or anyone for whom he is legally responsible interferes with progress of the works
- The employer or anyone for whom he is responsible fails to give the contractor possession of the site at the right time
- The employer suspends the works for one month
- The employer fails to comply with the CDM Regulations
- The employer becomes insolvent.

The contractor must follow the procedure precisely. The common law rights of both parties are preserved.

CHAPTER THIRTEEN
DISPUTE RESOLUTION

13.1 General

Since publication of the first edition of this book, a number of radical changes have taken place in the procedures available under the MW 80 for resolving disputes arising under the contract. No longer is arbitration the sole mechanism for dispute resolution and so it is necessary to consider this subject in a wider context.

Changes have come about by a combination of self regulation and the imposition of legislation. However, it is a sad reflection of the combative reputation of the industry that parliament has found it necessary to impose by far the most fundamental of those changes directly through the enactment of Part II of the Housing Grants, Construction and Regeneration Act 1996 (now commonly referred to as the Construction Act), and indirectly by enactment of the 1996 Arbitration Act.

In addition to the part that legislation has played, decisions in the courts are also continually bringing about the need for revision and refinement to the way in which contracting parties must review how they should regulate their contractual dispute resolution procedures. Over recent years numerous important judgments have significantly affected the drafting and the interpretation of the JCT contracts.

The relative simplicity of MW 80 has no doubt protected it from some of the effects of those decisions, such as that of the Court of Appeal in *Crown Estates Commissioners* v. *John Mowlem & Co Ltd* (1994) 70 BLR 1, which brought about significant re-drafting of the IFC 84 and JCT 80 conditions concerning the issue of final certificates under those contracts, but the 1998 decision of the House of Lords in *Beaufort Developments (NI) Limited* v. *Gilbert-Ash NI Limited and Ors* (1998) CILL 1386 will no doubt reach down even to MW 80. Thus, prospective parties to all of the JCT contracts, and perhaps especially MW 80, may now re-think their agreed procedure for resolving future differences.

Until the *Beaufort* case it was settled law that:

197

'in no circumstances would [the court] have power to revise a certificate or opinion [of the Architect] solely on the ground that the court would have reached a different conclusion from that of the architect giving the certificate or opinion since to do so would be to interfere with the agreement of the parties.'

In contrast, where the parties agreed to have their future disputes referred to arbitration, arbitrators appointed under MW 80 have express additional powers, conferred on them by the arbitration agreement, enabling them to open up, review and revise such certificates and opinions.

Since in the substantial majority of disputes under MW 80 the contractor will wish to go behind one or more of the architect's certificates and/or opinions, the decision in *Northern Regional Health Authority* v. *Derek Crouch Construction Co Ltd* (1984) 26 BLR 1 has for some 14 years militated in favour of arbitration as the preferred last resort method for settling construction disputes. In view of the *Beaufort* decision, it now remains to be seen whether that will continue to be the view, or whether things are set to change. In the meantime, parties to the MW 80 should, and in certain respects must, now address their minds to at least two distinct dispute resolution procedures where previously only the one remedy of arbitration was a viable option.

13.2   *Adjudication*

13.2.1   Introduction

This procedure is entirely new to the MW 80 Standard Form and, whether or not the parties wish to do so, they must now make provision for disputes and/or differences arising 'under' the contract to be referred to adjudication should either party wish to do so. The obligation to incorporate such a provision in the contract is a statutory one arising under Part II of the Housing Grants, Construction and Regeneration Act (commonly referred to as the Construction Act), enacted in 1996 to give legal effect to some of the proposals and recommendations put forward by Sir Michael Latham in his report of July 1994.

Although not expressly saying as much, Latham implicitly recommends that adjudication be available to provide parties to most 'construction contracts' with a quick and inexpensive means of initially resolving their otherwise insoluble disputes and/or differences. That recommendation is now translated into law via the Construction Act which, in turn, makes it mandatory for parties to those contracts to incorporate terms into their contract with the express objective of achieving those ends.

Contracts for the construction of private residences may escape the obligation. However, generally the obligation to incorporate such a provision is inescapable. The parties cannot contract out of that requirement and should they attempt to do so either wholly or partly, or should they otherwise neglect to comply with that requirement, they will in any event be bound to adopt statutory adjudication procedures laid down in subordinate legislation, known as the Scheme for Construction Contracts.

The purpose of the Act can briefly be summarised as imposing a legal obligation on the parties to expressly provide in their agreement for either party to give the other notice requiring any dispute or difference between them under the contract to be referred to an adjudicator within seven days of that notice. The contract must provide, too, for the matter concerned to be decided by the adjudicator within 28 days of that reference. That is subject to any agreement between the parties, after the reference has been made, to extend the time for a decision. The referring party may unilaterally extend the time for the decision to a maximum of 42 days. The adjudicator is to have wide powers in relation to the conduct of the proceedings and must, under the contract, be free to take the initiative in ascertaining both the facts and the law within the confines of absolute impartiality. Reaching his decision and making it known to the parties are clearly not one and the same thing.

Amendment MW11 of MW 80 has been drafted to provide 'ready made' terms satisfying those statutory requirements, thus avoiding the possibility that the parties need to rely on the 'fall back' procedure offered by the scheme. Conveniently, the amendments relating to the Construction Act are distinguished from others in Amendment 11 to assist parties who may wish only to incorporate those but none other of the changes brought about by Amendment 11.

13.2.2 Commencing the adjudication – notice to refer

Section 108(1) and section 108(2)(a) of the Act together require that the contract must provide that either party has the right, on giving notice, to refer a dispute arising under the contract for adjudication at any time. With the introduction of new Article 4.1 into MW 80, this right is now expressly translated into the contract so that the parties agree that any future disputes or differences which arise may be referred by either of them to adjudication, in accordance with the procedures set out in a new Part D of the Supplementary Memorandum.

Curiously, although section 108(1), and in turn Article 4.1, seem to suggest that *all* disputes and differences are capable of being referred to adjudication, those relating to the correct evaluation of VAT chargeable under clause B3.1 of the supplementary provisions are expressly excluded from the jurisdiction of an arbitrator appointed under the contract. Such disputes or differences must, if they are to be finally decided, be referred instead to the Commissioners of Customs and Excise.

Rights to arbitrate over disputes concerning the operation of the statutory tax deduction scheme, set out in part C of the supplementary provisions, are similarly excluded from the arbitration process, if and to the extent that any other Act or Regulation provides for some alternative method of resolving them. Why similar exclusions are not expressly made in respect of the adjudication process is not clear. Although the point may rarely be of any practical importance, one could envisage such an inconsistency leading to difficulties as the adjudication process becomes more widely used in time.

Little guidance is given under the contract or in the Act regarding what form the 'notice to refer' should take, but clause D4.1 does require that the notifying party must, as a minimum, briefly identify the nature of the matter being referred.

Although there may well be positive disadvantages in being overly explicit and excessively detailed at this early stage, it may be helpful if the notice was nevertheless sufficiently clear to alert those who may be called on to advise the employer or the contractor to the nature and extent of the dispute or difference concerned. Figure 13.1 shows a suitable letter.

13.2.3   Appointment of adjudicator

Speed is the key to the introduction, and more particularly to the eventual success or failure of the adjudication process. In an attempt to ensure the process is not frustrated by one or other reluctant party 'dragging their heels', the draftsmen of the Act laid down a strict and some would say unachievable timetable within which the process must be designed to operate.

Figure 13.1
Letter from one party to the other giving notice of referral

REGISTERED POST/RECORDED DELIVERY

Dear Sir

PROJECT TITLE

I/We hereby give notice that it is my/our intention to refer the following disputes or differences to adjudication in accordance with Article 4.1 of the contract between us dated [*insert date*].

The disputes or differences are:

[*Specify the disputes sufficiently so that the other party can readily identify them, but without setting out the detailed grounds of the case. For example: 'Failure of the architect to properly make an extension of time under the provisions of clause 2.2 of the contract'.*]

Yours faithfully

The contract need not stipulate a specific time limit within which the adjudicator's appointment must be made. It must, however, provide a timetable which has the object of both securing that appointment *and* referring the dispute to him within seven days of a notice requiring that adjudication be invoked.

Unfortunately, the notice provision (clause 1.5) is less than clear. Rather unhelpfully it suggests simply that any 'effective means' of service will suffice, but fails to explain whether or not the current (and in our opinion over used) trend for transmitting by facsimile, e-mails or other instantaneous electronic means, falls within that category. Whereas it might, of course, be argued that express reference in the clause to the recipient having to receive it at their agreed address implies postal service, that is by no means made clear. There is, however, little doubt that where there is no prior express agreement as to the relevant address for service, the clause then restricts the term 'effective means' to postal service. That then may give rise to considerable difficulty, not least because clause 1.5, (and indeed the Act at section 115), also rather unhelpfully provide that, if notice is to be deemed to have been effectively served, it must not only be addressed, pre-paid and posted, but must, according to clause 1.5, be 'delivered by post'. Because, where notices must be served by pre-paid letter, evidence of addressing and posting may prove insufficient and because, more importantly, in order to evidence delivery of a letter on a given date it may be relevant to show that the letter was, in fact, properly addressed, posted in due time and not afterwards returned, it is surprising that clause 1.5 does not at least suggest the use of recorded or registered post. One is left wondering how, in the event of any dispute over when or even whether the notice to refer was served in the first place, it would be resolved!

It can be seen, given the timetable with which the parties must comply, that securing the appointment of the adjudicator is a matter of considerable urgency. It is vital that the steps necessary and the time taken to select a suitable adjudicator, to invite him to take up the appointment, to agree his terms, and finally to 'secure' his appointment, must be kept to a minimum. There is no reason when the contract is made why the parties cannot agree who should in future adjudicate over any disputes or differences that may later arise. That clearly will speed up the appointment process considerably and such prior agreement is catered for in MW 80 both in the text of new article 4.1 and in Supplementary Memorandum D2.

However, any previously agreed nominee must first, according to provision D2.1 of the memorandum, agree with the parties to execute the JCT Standard Agreement for the Appointment of an Adjudicator, the text of which is conveniently reproduced in Amendment MW11. It follows that if the parties wish to preselect the adjudicator in this way, they must first consult with their prospective candidate to ensure that he is indeed willing to enter into that agreement and to act if called on to do so.

It cannot be disputed that there is a distinct advantage of speed where there is advance agreement on a suitable nomination. But that singular advantage must be weighed against other practical considerations. It may be that, if and when a future dispute arises, the proposed adjudicator finds himself unable to act at all or least at least within the stipulated period. He may, for example, have some conflict of interests or other prior commitment, or it may be that ill health or any number of other unforeseen events have arisen in the intervening period. Moreover, while the pre-agreed adjudicator may be entirely suitable and well qualified to consider and decide certain matters, such as proper valuation or defective design, he may be entirely ill equipped to properly consider a vast range of other potential issues that might subsequently arise. These may concern alleged defects in specialist workmanship and/or materials or issues relating to electrical, mechanical, structural or civil engineering design or workmanship. It is not difficult to imagine a situation where the parties' choice of candidate today would not later be their preferred choice in the light of the specific dispute as and when it arises.

If they do not agree on a suitable candidate – whether in advance or otherwise – or if for some other reason their previous nominee cannot or will not act, the contract offers a choice of bodies ('nominating bodies') to whom either party may unilaterally apply for an appropriate appointment. They are:

- The Royal Institute of British Architects
- The Royal Institution of Chartered Surveyors
- The Construction Confederation
- The National Specialist Contractors Council.

From that list the parties should, at the time of contracting, choose one, otherwise the task will by default fall to be carried out by the Royal Institute of British Architects.

Whichever nominating body is selected, clause D2.1 again makes it clear that the prospective nominee must undertake to execute the standard agreement for his appointment issued by the JCT. Because, by footnote [†] to Part D it is confirmed that the nominating bodies concerned have agreed they will nominate only those individuals known to be prepared to do so, the parties can be satisfied that the nominated bodies listed will ensure that will be the case.

Clause D2 of MW 80 makes it clear that at whatever stage and by whatever means the adjudicator is appointed, he/she acts in an individual capacity. That is to say, the task cannot be undertaken by a corporate body such as an architectural or quantity surveying practice, law firm or other consultancy. What is more, however the appointment is secured, clause D2.2 emphasises the parties' obligation to take all practicable steps to secure the appointment without delay, such that the further timetable for reference of the dispute to the adjudicator and the eventual publication of his decision can be achieved. Not to do so will be a breach of contract.

13.2.4   Referral of the dispute

Once the adjudicator is appointed, it is for the party responsible for invoking the process to refer the dispute or difference to the chosen adjudicator and in making 'the referral' to do so by the later of:

- seven days from the date of the initial notice to refer
- seven days from execution by the parties and the adjudicator of the JCT Adjudication Agreement.

With due respect to the draftsman, while the logic of linking 'the referral' to execution of the JCT Adjudication Agreement by the parties and by the adjudicator cannot be denied, it is difficult to see how it can be said that this alternative timing satisfies the single obligation laid down in the Act to:

'provide a timetable with the object of securing the appointment of the adjudicator and referral of the dispute to him within 7 days of such notice'.

As and when 'the referral' is made it must simultaneously be copied, in its entirety, to the other party and in both instances it must include with it:

- Particulars of the dispute or difference
- A summary statement of the contentions relied upon
- A statement of what remedy or relief is being sought

- Any material that the referring party wishes the adjudicator to consider.

Unlike the prior notice referring the dispute to adjudication and service of other notices and/or documents to which clause 1.5 applies, to be effectively served the referral must be either hand delivered, sent by recorded or registered post, or rather surprisingly can be served by fax. In the latter case the referral (including all accompanying documentation) must then be sent 'for record purposes' forthwith, either by first class mail or by actual delivery.

On a strict reading of clause D4.2, it appears the use of recorded or registered post although mandatory if the referral is first sent by post, is not so where it was first sent by facsimile. Where faxed in the first instance, the clause seemingly allows a subsequently posted copy to simply be served by ordinary first class post, since clause D4.2 makes it clear that subsequent posting of a prior facsimile is taken as merely 'for record purposes only'.

It follows that, if served by fax, the date (and time) of referral will be taken (subject to proof if required) as when the facsimile could reasonably have been expected to have been received and read by the intended recipient. Under MW 80, it seems therefore that the referral can be made instantaneously, so allowing it to be sent on the last hour of the seventh day following which notice to refer was given. That said, reliance on that facility when setting one's timetable in readiness for adjudication should be treated with caution, bearing in mind that whether sent by fax or not the referral must be accompanied by 'all' of the documents relied upon. In the case of a dispute involving photographic evidence, drawings and/or many files of papers, faxing may well prove impossible or impracticable as a proper means of service.

Hence, although it may appear an attractive proposition where time is tight, the contract does not contemplate, nor does it expressly sanction, referral of some documents by fax and others by post. Any attempt to adopt such a procedure would, it is suggested, make the use of the fax redundant and so the date of 'the referral' would then, in any event, be taken as the date of receipt (or deemed date of receipt) of the posted material and may then be out of time.

However sent, the adjudicator should, and according to clause D5.1 of the supplementary memorandum must, confirm to both parties that he has received the referral and its accompanying documents. Interestingly, unlike the referral (and, indeed, the statement in response discussed below), that confirmation is not expressly sanctioned to be sent by fax.

Once the referral has been made and received, the non referring party has an express right of reply. That reply must similarly be given in writing and must again both set out the contentions relied upon and be accompanied by any supporting material. The mode and manner of service of this statement, both to the adjudicator and to the other party, are once more governed by the conditions (in clause D4.2) governing the referral of the dispute, as discussed above.

There is no doubt that the procedure laid down in the contract requiring each party to provide written statements of their position is eminently sensible. There is no doubt, too, that that process reflects what would otherwise be the most commonly adopted procedure. However, it is surprising to note that this contractual procedure does not, on the face of it, strictly reflect the provisions of the Act.

Whereas the Act requires merely that the referral be made within seven days of notice to refer, it does not stipulate either the form and/or contents of that referral. Nor is there any stipulation in the Act concerning service of any reply by the non referring party. It would seem that that omission on the part of the draftsman is quite deliberate since the Act provides, (at section 108(2)(f), that it is the adjudicator who must be empowered to take the initiative in determining both the facts and the law.

Whilst under the contract (clause D5.5) the adjudicator is given almost unfettered discretion, for the purposes of reaching his decision, to set down his own procedure for conduct of the adjudication, he and the parties are seemingly bound in the first instance, under MW 80, to deal with the matter according to a predetermined procedure of written submissions, along the lines of the documents-only process under the JCT Arbitration Rules. In that case, unless the adjudicator attempts to override the contractually agreed mechanism of clauses D4.2 and D5.2 respectively (and it is by no means certain he can), and if the time limits of those clauses are utilised to the full, then it follows that no decision can even begin to be reached until seven days after the non-referring party has received confirmation that the referral has been received by the adjudicator. In effect, the decision-making process may possibly not even begin until 14 days after the notice requiring the dispute to be adjudicated was first given by the aggrieved.

Albeit that the adjudicator has express power to set his own procedures and more fundamentally may take the initiative in order to ascertain whatever facts and legal questions he considers are necessary, there are two basic parameters within which he must work and, indeed, must clearly be seen to work. Both under the Act and more expressly under the contract (by reference to clause D5.5) he must ensure that his decision is reached with absolute impartiality and he must ensure that he keeps within the relative confines of the matters referred to him. The jurisdictional line might be drawn very narrowly: *Cameron* v. *John Mowlem* (1990) 52 BLR 24. However, subject to those two parameters the adjudicator's powers are significant. They expressly extend to:

- Using his own knowledge and/or experience
- Opening up, reviewing and revising any certificate, opinion, decision, requirement or notice issued, given or made under the contract
- Requiring further information than that contained in the notice of referral and its accompanying documentation or in any written statement provided by the parties, including the results of any tests or of any opening up
- Requiring the parties to carry out tests or to open up work
- Visiting the site of the works or any workshop where work is prepared for the contract
- After notice, obtaining such information as he considers necessary from any employee or representatives of the parties
- After notice and estimate, obtaining from others such information and advice as he considers necessary on technical and legal matters.

Subject to taking into account any contractual provision in that regard, such as the provision in MW 80 concerning interest on late payment, the adjudicator also has discretion to decide in what circumstances and for what periods he may order a paying party to pay interest to the other. Whilst the drafting of the clause might lead some to insist that the adjudicator is bound to apply simple interest at the statutory rate to which the courts are generally constrained, it seems likely that the decision on what rate shall apply is left, too, to the impartial discretion of the adjudicator.

The thorny subject of costs is dealt with succinctly, with both parties being responsible for their own costs of the proceedings, irrespective of the outcome. Those with first hand experience of the significant 'tactical' part that those costs play in the more traditional dispute resolution procedures, whether it be arbitration or litigation, will surely see this innovation as offering adjudication a real chance of achieving its ultimate aim of promoting quick, economical and if possible informal resolution of genuine disputes. The corollary is of course the creation of a disincentive for parties to generate spurious or at the very least speculative disputes. The chance of success often turns on the other party's fear and perceived risk, however strong they otherwise think their defence might be, that they might ultimately be liable for not only their own costs but those of their opponent. The risk is not entirely removed since the adjudicator is free to decide how his fees, and those of any testing or the like that he directs should take place, should be apportioned. But again, the contract recognises the possibility that those, too, might ultimately be borne in equal proportions irrespective of the outcome. Indeed, that will be the case in the event that the adjudicator does not expressly direct otherwise, but it should be remembered that the parties remain jointly and severally liable for those fees and reasonable expenses incurred by the adjudicator 'pursuant to the Adjudication'.

Once made, the adjudicator's decision is binding unless and until the relevant dispute is finally decided by agreement, or is decided either by an arbitrator appointed in accordance with the arbitration agreement, or by the courts if appropriate.

In keeping with the underlying philosophy of adjudication, the contract makes it clear that any reference, either to an arbitrator or the courts, of a dispute previously referred to adjudication is not an appeal against the adjudicator's decision. It will be a complete reconsideration of the dispute, conducted in the normal way in accordance with the Arbitration Act and any agreed Arbitration Rules or the full judicial process.

13.3 Arbitration

13.3.1 Introduction

Following the decision in *Beaufort*, it remains to be seen whether the long standing use of arbitration will continue to be the normal mode by which construction industry disputes are finally resolved. Anticipating that it will, MW 80, in common with all other JCT contracts, has undergone an extensive revision of the arbitration provisions to reflect the significant overhauling of the arbitration process brought about both by the latest (1996) Arbitration Act and by the introduction of the JCT 1998 Edition of the Construction Industry Model Arbitration Rules (CIMAR), which now replace, under MW 80, the earlier use of the JCT Arbitration Rules which were themselves the cause of the last major review of the arbitration provisions of this contract. The provisions for arbitration are now set out in a new Article 4.2 and in a new Supplementary Memorandum Part E, both of which are introduced into the MW 80.

Disputes and/or differences arising under or connected with the contract between contractor and employer, or the architect on his behalf (except those relating to VAT and to the statutory tax deduction scheme where those are catered for under statute), must be submitted to arbitration. However, with the introduction by Article 4.1 of the concept of adjudication, reference to arbitration is now also made subject to the parties' right to first invoke that interim adjudication process. A further important proviso is also made so that the 'enforcement' of any such prior decision by an adjudicator is now added to the short list of exceptions to the otherwise unlimited scope of the matters which the parties can (and indeed must) have arbitrated. Although not, therefore, the first available means of dispute resolution, arbitration remains, for all material purposes, the last resort and as such the parties and architect alike should do all they can to avoid it. But if arbitration is threatened and if it should prove impossible to properly resolve it informally, the architect must grasp the nettle.

The eventual outcome of arbitration is always uncertain and there has been a growing trend in recent years for claimants to use the risk, however small, of an adverse decision which carries the associated risk of liability for the parties' and the arbitrator's combined costs, to force an offer of settlement in what might otherwise be a speculative and unmeritorious claim. That approach is unlikely to disappear overnight simply because of the introduction of a requirement for the parties to first have a right to adjudicate their dispute. Employers, architects and contractors alike must therefore appreciate how the arbitration process will operate so that, at the very least, they might fully appreciate the possible consequences of formal proceedings.

For all its technical differences, in essence, arbitration should be seen as nothing short of formal legal proceedings. A common misconception among those embarking for the first time along the long and often tortuous arbitration route is that the process will have an air of informality about it, akin to a debate during which each party simply argues out their position on an ad hoc basis before the arbitrator in a rather casual manner. The arbitrator, they imagine, takes an inquisitorial role, testing what each party has to say and inviting comment and response from the contestants as to what they have to say about the other's point of view concerning their own and their opponent's case. Whether or not that would, in fact, prove to be a more satisfactory approach is a moot point and no doubt arbitration (and indeed our legal system) evolved from just such a process.

However, the system today, certainly so far as it is applied to the construction industry, is a far cry from that rather informal and inquisitorial approach. Like the judicial system, arbitration is adversarial. However, it offers a degree of flexibility beyond that available to litigants, with the disputants in arbitration being free to agree on who should act as their arbitrator, or at the least agreeing on a third party with power to make that appointment. They have freedom, too, to agree important matters such as the timetable of the proceedings – with the possibility of a quicker procedure than would be the case in litigation – and the venue to suit the convenience of the parties and their witnesses.

Another important feature is that of confidentiality, with the hearing being conducted in private with the parties being free to be represented other than by a solicitor and/or counsel. Often, nowadays, they are represented by experts with dual professional and legal qualifications who have long and continuing experience as architects, engineers or quantity surveyors or the like in the industry, and who also have considerable specialist knowledge and experience of acting in the care and conduct of arbitration proceedings.

13.3.2 Procedure

Since the introduction of the JCT Arbitration Rules in 1988 there has been a tailor-made procedure for conducting and settling construction arbitrations under the JCT standard form contracts should the parties wish to adopt it.

Although not beyond criticism, the rules have generally provided parties with a pre-agreed procedure and framework within which their arbitration proceedings must be conducted, and to that extent if no other, they remove the opportunity for further argument and time wasting debate over purely procedural matters. With the radically different approach to such rules dictated by the introduction of the 1996 Arbitration Act, and following extensive consultation with the industry over something in excess of 18 months, the JCT Arbitration Rules have now been replaced by the Construction Industry Model Arbitration Rules (CIMAR) whose stated objectives are to extend or amend the provisions of the Act where necessary and, in a 'user friendly' way, to briefly but clearly add a general framework to the specific powers and duties in order to provide guidance to users and arbitrators alike.

If a pre-Amendment 11 edition of the standard MW 80 is used, any arbitration under the contract will be conducted according to the JCT Arbitration Rules unless the parties expressly provide otherwise. A footnote to the relevant provisions provides the parties with a 'health warning', drawing attention to the fact that:

> 'The JCT Arbitration Rules contain stricter time limits than those prescribed by some arbitration rules or those frequently observed in practice. The parties should note that a failure by a party or the agent of a party to comply with the time limits incorporated in these Rules may have adverse consequences.'

A second footnote invites the parties to delete clause 9.5 if they do not want the JCT rules to apply. (For a useful overview of the use of the JCT rules see the previous edition of this book.)

Use of the post Amendment 11 edition of MW 80, on the other hand, provides that arbitrations are to be conducted in accordance with the CIMAR rules. Unlike the pre-Amendment 11 edition the parties are not discouraged by any footnote from contracting out of the use of the rules. Indeed, to avoid their use will now involve wholesale amendment and/or deletion of substantially the entire Supplementary Memorandum Part E. Even more so than with the old JCT rules, there can be no valid reason for wanting to avoid the use of CIMAR and, indeed, there are many good reasons for conducting the arbitration in accordance with them. In particular these more recent rules retain a choice of three procedures for the conduct of the proceedings: short hearing procedure, documents only procedure and full procedure.

Short hearing procedure

This is a useful and sensible procedure. Subject only to any contrary agreement between them, it gives the parties no more than one day during which they will both have reasonable opportunity to address the matters in dispute orally before the arbitrator. Before then, both parties will have either simultaneously or sequentially submitted written statements of their case, along with any accompanying documents and witness statements on which they rely, to the arbitrator. The arbitrator either before or after the short hearing will have the opportunity in appropriate circumstances to inspect the subject matter of the dispute.

It is a procedure particularly well suited to short disputes, capable of determination principally by means of inspection of work, materials, plant or equipment and the like, and where the arbitrator will generally be able to decide the issues and make his award within a month or so of hearing each party's representations. The parties are free to present expert evidence to support their case. But this costly (and often unnecessary process, where the arbitrator is often selected largely because of his own specialist expertise in the particular field in dispute), is to some extent discouraged by the rules.

Documents only procedure

Again a useful and much under-used procedure, this is appropriate where the dispute can properly be disposed of without the need for oral evidence or alternatively, and as importantly, where the sums in issue do not warrant the time, and hence the additional expense, of a hearing.

This is a procedure that warrants consideration particularly in the context of the nature and relative size of disputes that might arise under MW 80. Each party either simultaneously or sequentially, as the arbitrator directs, will serve on each other and to the arbitrator a written statement of their case which will include with it, at least, an account of the relevant facts and/or opinions relied upon, relevant witnesses' statements concerning those facts or opinions, and a statement of the relief or remedy that is being sought. There is provision for each party to reply to the others' statement(s) and for the arbitrator to then put questions or to ask for further statements as he considers appropriate. There is an ultimate proviso that the arbitrator may decide that a day be set aside for him to question the parties and/or their witnesses and for the parties to comment on any additional information that may have been provided.

Given the nature of the disputes that will fall into this category of procedure it is anticipated that, like the short hearing procedure, the arbitrator should be able to reach his decision within a month or so of the parties' final exchanges and any subsequent questioning.

Full procedure

This is useful where neither of the preceding two options is considered to be satisfactory and where, for example, there is also significant disagreement as to the facts and/or the opinion evidence of experts such that oral evidence subject to cross examination would be appropriate. The rules, and in particular rule 9, provide for the dispute to be dealt with and the matters decided using procedures akin to those conventionally adopted in the High Court.

Since arbitration like litigation must be capable of providing a final resolution of disputes, ranging from simple issues over hundreds or thousands of pounds to complex disputes involving hundreds of thousands or even millions, the full procedure must, despite the rules, be capable of modification in the same way as High Court proceedings, to cope with that potential complexity in both a cost efficient and timely way. The rules therefore provide a framework around which the arbitrator is free to:

> 'permit or direct the parties at any stage to amend expand, summarise or reproduce in some other format any statement of claim or defence so as to identify the matters essentially in dispute including preparing a list of matters in issue.'

Subject to that discretion, the full procedure, if adopted, requires the parties to exchange formal statements (or pleadings) comprising respective statements of claim and their corresponding defence by which they will each provide the other with sufficient particulars to enable the other to answer each allegation.

To that end, each statement should:

- Contain the facts and matters of opinion which are intended to be established by evidence and may include a statement of any relevant point of law
- Include or refer to evidence to be presented if this will assist in defining the issues
- State the relief or remedies sought, for example specific monetary losses, in such a way that they can be answered or admitted
- Adopt a common system of numbering or identification of sections to facilitate analysis of the issues
- Anticipate the need to incorporate replies where particulars are given in schedule form.

With those basic requirements in mind, the arbitrator must then give detailed directions as to the timing for the preparation and service of the respective statements and for all other steps that are likely to be encountered prior to the hearing, such as the provision of any further and better particulars of either party's earlier statement, disclosure by the parties of any documents material to the issues, exchange of any statements of witnesses of fact, any arrangements for the preparation and exchange of experts' reports, and arrangements for hearing the parties (or their advocates) and the respective witnesses.

13.4   *Appointing an arbitrator*

Under MW 80, arbitration is possible during the currency of the contract. There is no need to wait until the works are completed or the contractor's employment has been brought to an end. All that is needed is for there to be 'any dispute or difference as to any matter or thing of whatsoever nature arising under this Agreement or in connection therewith', excepting anything connected with the enforcement of any adjudicator's decision, or disputes, or VAT, or statutory tax deduction scheme issues.

Either party can then set the procedure in motion, and the relevant arbitration proceedings will begin in respect of a dispute when one party serves written notice on the other, in the manner provided by rule 2.1 of CIMAR, identifying the dispute concerned and requiring agreement to the appointment of a suggested arbitrator.

In inviting agreement to the appointment of an arbitrator, it is always best to suggest the names of three people, any of whom would be suitably qualified and acceptable to act. Beyond the obvious requirement for the nominees to be competent, experienced and suitably qualified to act, it is essential too that they be independent and have no connection with or interest in the parties or any matter connected with the dispute. Figure 13.2 is a suitable letter.

The Chartered Institute of Arbitrators, International Arbitration Centre, 24 Angel Gate, City Road, London EC1V 2RS (Telephone 0171 837 4483) is always willing to suggest the names of suitable people on request. But an important aspect of any choice is a proper understanding of the nature of the dispute and the sums involved, and the architect should ensure that these are clearly appreciated from the outset.

No matter how well experienced or eminently suitable the suggested nominations might be, it may well be (and regrettably without good cause it is all too often the case) that the other party will write back rejecting the proposals and suggesting other names instead. Those should not in turn be rejected out of hand since agreeing on an arbitrator is far better than leaving the appointment in the hands of a third party.

However he or she is selected, the arbitrator must be:

Figure 13.2
Letter from one party to the other requesting
concurrence in the appointment of an arbitrator

REGISTERED POST/RECORDED DELIVERY

Dear Sir

PROJECT TITLE

I/We hereby give you notice that we require the undermentioned dispute(s) or difference(s) between us to be referred to arbitration in accordance with Article 4.2 of the contract between us dated [*insert date*]. Please treat this as a request to concur in the appointment of an arbitrator under clause E2.1.

The dispute(s) or difference(s) is/are:

[*Specify*]

I/We propose the following three persons for your consideration and require your concurrence in the appointment of one of them within 14 days of the date of service of this letter, failing which I/we shall apply to the President of the [Royal Institute of British Architects *or* Royal Institution of Chartered Surveyors as appropriate] for the appointment of an arbitrator under Article 4.2.

The names and addresses of the persons we propose are:

(1) Sir Bertram Twitchett, RIBA, FCIArb, Fawlty Towers, Probity, Hants. *etc.*

Yours faithfully

- Independent
- Impartial – not having any existing relationships with the architect, the employer, the contractor or anyone else involved
- Technically and/or legally qualified as appropriate.

In practice, of course, if it has become necessary to request arbitration, relationships between those involved have gone sour and requests to concur in an appointment are often ignored or rejected out of hand. The contract therefore provides that, if the parties fail to agree an appointment within 14 days of the written request (or within any agreed extension to that time), the arbitrator will be appointed by a third party – the President or a Vice-President of the Royal Institute of British Architects, the Royal Institution of Chartered Surveyors or the Chartered Institute of Arbitrators. Article 4.2 should have been completed so as to delete one or other of these appointing bodies, but if it has not then the President or a Vice-President of the Royal Institute of British Architects will make the appointment on written application to do so by either party. Figure 13.3 is the sort of letter that a contracting party might send to an appointing body requesting appointment. However, it will be necessary to complete special forms and to pay a fee to the appointing body.

It should be closely noted that both MW 80 and the rules make it clear that the proceedings are begun then only in respect of the particular dispute to which the notice refers. Rule 3.6 of CIMAR provides further that arbitral proceedings in respect of any other dispute are begun when notice of arbitration for that other dispute is served. Since, commonly, recipients of such a notice will no doubt want, in due course, to raise a counterclaim, that drafting raises a number of questions the answers to which are not, it must be said, immediately apparent from either the contract or the rules.

Arbitrators do not work for nothing; they are professionals and charge professional fees. Although there is no fixed scale of fees, most arbitrators charge between £800 and £1,600 a day, the actual fee charged being based on the standing and expertise of the arbitrator and the complexity of the dispute. Whether or not terms have been agreed, the arbitrator's appointment, if made by consensus, will take effect upon him signifying his agreement to act. If made by the Royal Institute of British Architects or other agreed nominating body, the appointment becomes effective, again whether or not terms have been agreed, when the appointment is made by the relevant body.

Figure 13.3
Letter from architect on employer's behalf or from contractor to appointing body

---

Arbitration Appointments Office,   The Appointments Secretary,
Royal Institute of   (Arbitrations),
British Architects,   Royal Institute of
66 Portland Place,   Chartered Surveyors,
London W1N 4AD   Great George Street,
     London SW1

Dear Sir

PROJECT TITLE

[I am acting as architect for [*name of employers*] *or* We are the contractor] under a contract in MW 80 Form, Article 4.2 of which makes provision for your President to appoint an arbitrator in default of agreement.

Will you please send me/us the appropriate form of application and supporting documentation, together with a note of the current fees payable on application.

Yours faithfully

13.5 *Progress and powers of the arbitrator*

Once his appointment is made and becomes effective, the arbitrator will consider which of the forms of procedure appears to be most appropriate to the circumstances of the dispute and will enable the parties to both put their own case and deal with that of their opponent. The dispute must also, however, be kept in context, and in setting the procedure the arbitrator will also have an eye to adopting the format that will best avoid undue cost and/or delay. For even the most experienced of arbitrators that can often be a difficult balance to strike and so the parties must, as soon as possible after his appointment, provide him with an outline of the dispute and of the sums in issue and an indication of which procedure the parties themselves consider might be best suited to the dispute.

After considering the nature of the dispute and after taking account of their views on the matter, the arbitrator will generally arrange to meet with the parties and/or their representatives and will instruct them concerning which procedure is to be adopted and how, if at all, it should be tailored, whether in relation to the imposition of any particular time limitations or otherwise, to suit the specific reference.

Although it is always open to a party to conduct their own case, if the dispute has reached this stage we suggest that it should be handed over to consultants with particular expertise in the care and conduct of arbitration proceedings.

The arbitrator derives his powers from the arbitration agreement itself (Article 4.2 and Supplementary Memorandum Part E) and from the Arbitration Act 1996. Under clause E3 of part E (subject to the provisions of Article 4.2) the arbitrator, in addition and without prejudice to his general powers, has express power to rectify the contract – that is to say to correct written errors in it where it fails to represent what the parties agreed. He can also effectively substitute his own decisions for those of the architect when he exercises his power to 'open up, review and revise any certificate, opinion, requirement or notice'.

It is important to be aware that, unless the parties have already agreed in what circumstances the authority of the arbitrator may be revoked, or otherwise act jointly in writing to do so, subject only to very limited circumstances and/or to the powers of the High Court to so order, his authority is irrevocable. It is also important to understand the lengths to which that authority and power may extend under the Act. By way of example (and subject to any specific agreement between the parties) an arbitrator may:

- Order whether and if so which documents or classes of documents should be disclosed between and produced by the parties (section 34(2)(d))
- Decide whether and to what extent he should take the initiative in ascertaining the facts and the law (section 34(2)(g))
- Order whether the strict rules of evidence, or any other rules, as to the admissibility, relevance or weight of any oral, written or other material which either party may wish to produce, shall apply. (section 34(2)(f))
- In appropriate circumstances dismiss the claim or otherwise continue the proceedings and make an award based on the information before him in the absence of a party or without any submissions on its behalf in the event of inexusable delay or other failure to properly and expeditiously conduct the arbitration (section 41(1) to section 41(4))
- Make more than one award at different times on different aspects of the matters to be determined (section 47(1))
- Direct (and/or timeously review any previous direction) that the recoverable costs of the arbitration, or any part of the arbitral proceedings, shall be limited to a specified amount (section 65(1) and section 65(2))
- Make an award allocating liability for the recoverable costs of the arbitration as between the parties (section 61(1) and section 61(2))
- Award interest (section 49(1) to section 49(6))
- Take legal or technical assistance or advice (section 37)
- Order security for costs (section 38)
- Give directions as to the inspection, photographing, preservation, custody or detention of any property owned by or in the possession of any party to the proceedings, which is the subject of the proceedings (section 38).

Figure 13.4 shows the adjudication and arbitration in basic outline.

13.6 Summary

- MW Article 4.2 states that, subject only to the very limited exceptions of issues relating to VAT, the statutory tax deduction scheme and 'enforcement' of an adjudicator's award), arbitration is to be the procedure for finally resolving disputes and/or differences arising under or connected with the contract.

Figure 13.4
Flowchart of adjudication/arbitration

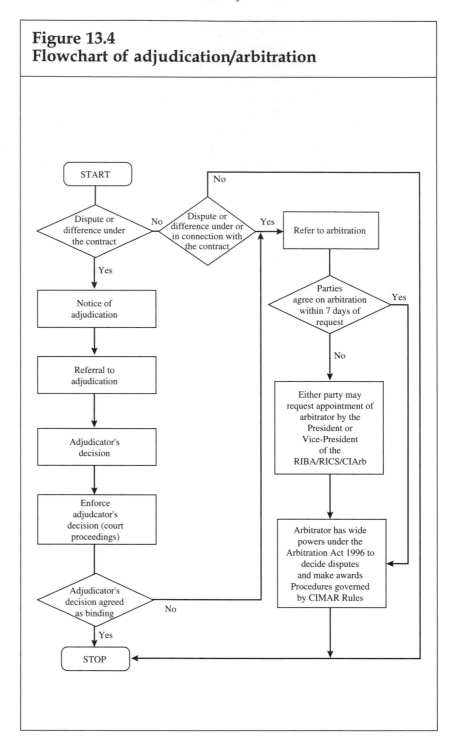

- Arbitration can be commenced by either party at any time.
- Failing agreement between the parties on a suitable arbitrator, unilateral application can be made to – and a suitable appointment will be made by – one or other of the pre-agreed nominating bodies named in the contract, namely either the RIBA, RICS or Chartered Institute of Arbitrators.
- Arbitration under the post-Amendment 11 edition of MW 80 will be conducted under and be subject to the CIMAR Rules. Those under the pre-Amendment 11 edition will be conducted subject to and in accordance with the JCT Rules.
- Whether the pre- or post-Amendment 11 is used, the Arbitration Act 1996 will apply.
- Where there is an effective arbitration agreement there is, under the Act, an effective veto in one or other party opting to litigate the dispute.
- The arbitrator has extremely wide powers.
- If at all possible, both litigation and arbitration should be avoided.

Table of Cases

Clause Number Index to Text

Subject Index